Autorenkollektiv

Structural Chemistry

Autorenkollektiv

Structural Chemistry

With 72 Figures and 11 Tables

AKADEMIE-VERLAG · BERLIN
1984

Die Originalausgabe erscheint im Springer-Verlag
Berlin · Heidelberg · New York
in der Schriftenreihe „Topics in Current Chemistry", Volume 123

Vertrieb ausschließlich für die DDR und die sozialistischen Länder

Erschienen im Akademie-Verlag, DDR-1086 Berlin, Leipziger Straße 3—4
Lizenznummer: 202 · 100/523/84
Printed in the German Democratic Republic
Gesamtherstellung: VEB Druckerei „Thomas Müntzer", 5820 Bad Langensalza
Umschlaggestaltung: Karl Salzbrunn
LSV 1215, 1275
Bestellnummer: 763 395 6 (6846)

08800

Table of Contents

The Calixarenes

C. David Gutsche

Department of Chemistry, Washington University, St. Louis, Mo., 63130, USA

Table of Contents

1 Introduction

The synthesis of compounds containing cavities of molecule-sized dimensions has captured the attention of numerous chemists in recent years, and this area of organic chemistry is acquiring the status of a recognizable and expanding subdiscipline. The principal reason for the burgeoning interest in these compounds is their imputed, and in some instances demonstrated, ability to form inclusion complexes, *i.e.* to participate in what has been variously described as "host-guest" chemistry [1] or "receptor-substrate" chemistry [2]. This review deals with certain $[1_n]$metacyclophanes possessing basket-like shapes, particular attention being given to those members which have been named "calixarenes" [3,4].

2 Nomenclature of the Calixarenes

The compounds discussed in this review are represented by the general structure *1*. In the IUPAC system of nomenclature [5], a specific member of this group (as represented by structure *2*) is named

pentacyclo[19.3.1.$1^{3,7}$.$1^{9,13}$.$1^{15,19}$]octacosa-1(25),3,5,7(28),
9,11,13(27),15,17,19(26),21,23-dodecaene ,

and it is numbered as shown in Fig. 1. An alternative nomenclature for this type of ring structure was suggested by Cram and Steinberg [6,7] according to which *2* is named as [1.1.1.1]metacyclophane. Several research groups have reported syntheses of the tetrahydroxy derivatives of *2* (as represented by structure *3*) and have named them in various ways. Zinke and coworkers [8] called these compounds "cyclischen Mehrkernmethylene-phenolverbindungen", Hayes and Hunter [9] named them "cyclic tetranuclear novolaks", and Cornforth and coworkers [10] referred to them as "tetrahydroxycyclotetra-*m*-benzylenes. For convenience of written and verbal discussion

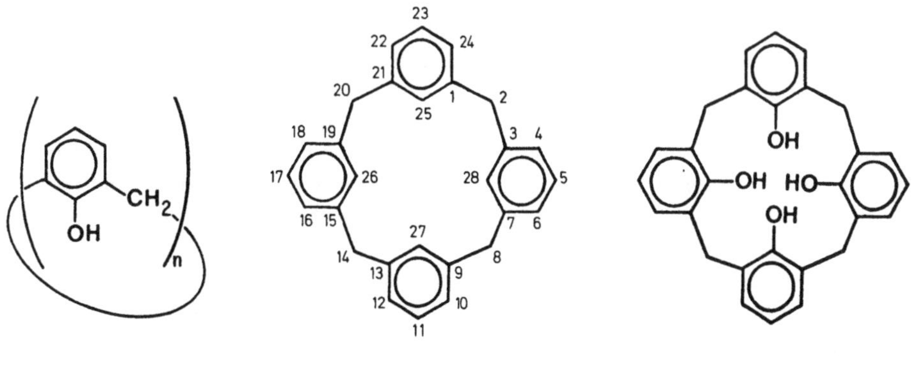

1 *2* *3*

Fig. 1. Poly-aryl *m*-methylene-bridged macrocyclic compounds we have chosen to call them "calixarenes" (Greek, *calix*, chalice; arene, indicating the incorporation of aromatic rings in the macrocyclic array), specifying the size of the macrocycle by a bracketed number inserted between *calix* and *arene* and specifying the nature and position of substitution on the aromatic rings by appropriate numbers and descriptors [11].

The structures and numbering for five types of calixarenes containing intraannular hydroxyl groups [12] which figure prominently in the following sections of this review are shown in Fig. 2. The cyclic tetramer composed of *p-tert*-butylphenol units and methylene units, for example, is named 5,11,17,23-tetra-*tert*-butyl-25,26,27,28-tetrahydroxycalix[4]arene; in abbreviated fashion it will be referred to as *p-tert*-butylcalix[4]arene.

3 Synthesis and Characterization of the Calixarenes

3.1 Arene-Aldehyde Condensations

3.1.1 Base-Catalyzed Condensation of *p*-Substituted Phenols and Formaldehyde

In 1872 Baeyer heated aqueous formaldehyde with phenol and observed a reaction which yielded a hard, resinous, noncrystalline product [13]. The chemical techniques at the time were not sufficiently advanced to allow characterization of such materials, however, and the structure remained unknown. Three decades later Baekeland devised a process for using this phenol-formaldehyde reaction to make a tough, resiliant resin (called a phenoplast) which he marketed under the name "Bakelite" [14] with tremendous commercial success. As a result, considerable industrial and academic attention was focused on phenol-formaldehyde processes, and a significant literature arose dealing with phenoplasts. Among these investigations were ones carried out by Zinke and coworkers in connection with the "curing" phase of the process [8,15–19]. In the investigation of this phenomenon they treated various *p*-substituted phenols with aqueous formaldehyde and sodium hydroxide, first at 50–55 °C, then at 110–120 °C for 2 hours and, finally, in a suspension of linseed oil at 200 °C for several hours. From *p*-methyl, *p-tert*-butyl, *p-tert*-amyl, *p*-(1,1,3,3-tetramethylbutyl), *p*-cyclohexyl, *p*-benzyl, and *p*-phenylphenol very high-melting, highly insoluble materials were obtained, all of which were postulated to be cyclic tetramers, *i.e.* calix[4]arenes of structure *4* in Fig. 2. The tacit assumption that a single product is formed in every instance was later shown to be incorrect by Cornforth and coworkers [10] who isolated higher- and lower-melting compounds from the condensations of formaldehyde with *p-tert*-butylphenol and *p*-(1,1,3,3-tetramethylbutyl)phenol (often referred to in the calixarene literature as *p*-octylphenol). Cornforth's conclusion that the materials were conformational isomers of the calix[4]arenes, however, was subsequently invalidated by Kämmerer and coworkers [20,21] and by Munch [22] whose temperature dependent ^{1}H NMR studies showed that rapid conformational interconversion occurs at room temperature. Finally, the recent work of Gutsche and coworkers [23] has revealed that mixtures comprising cyclic oligomers of *various ring size* are generally obtained in these condensation reactions. In the most thoroughly studied example [3,23–28] it has been shown that the condensation of *p-tert*-butyl-

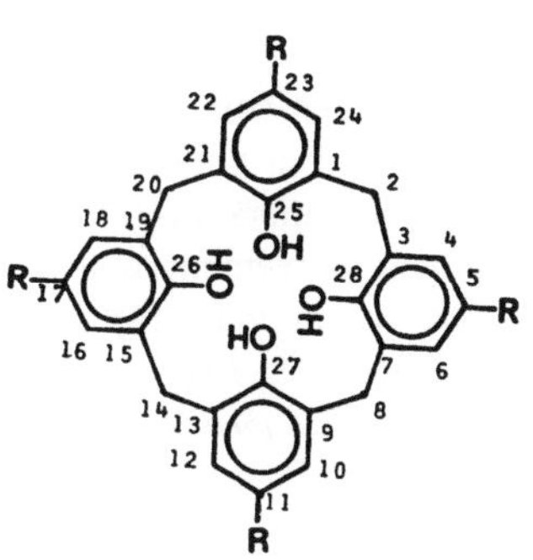

4

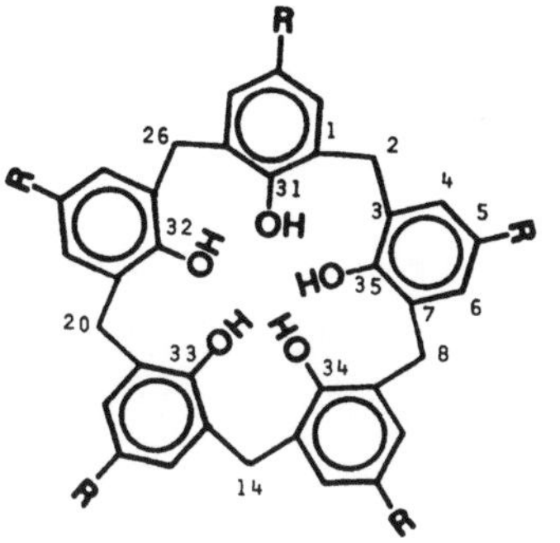

5

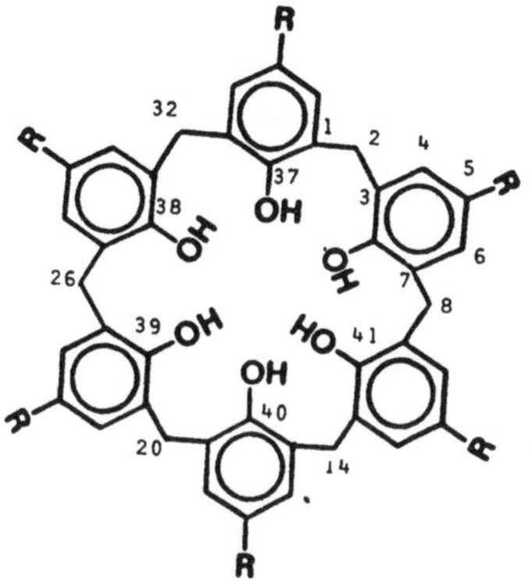

6

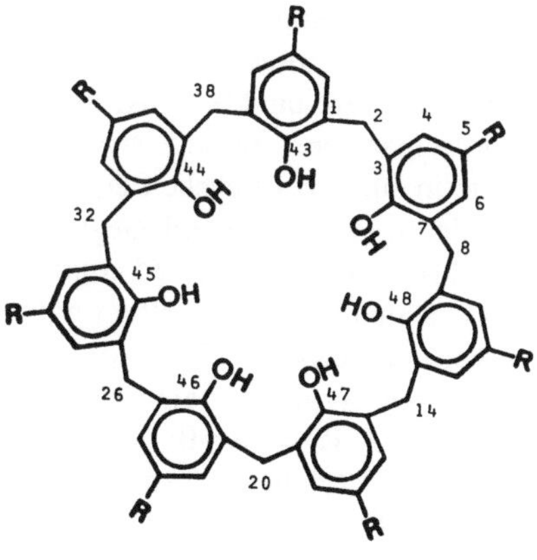

7

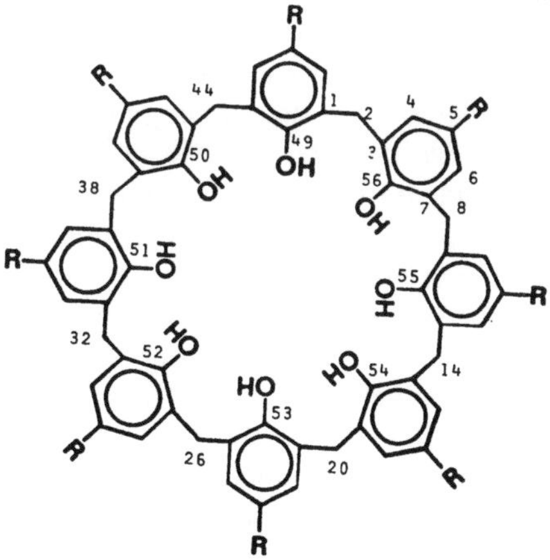

8

Fig. 2. Structures and numbering of intraannularly-hydroxylated calix[n]arenes phenol and formaldehyde yields cyclic tetramer (*4*, R = *tert*-Butyl), cyclic hexamer (*6*, R = *tert*-Butyl), and cyclic octamer (*8*, R = *tert*-Butyl) as the major products as well as small amounts of cyclic pentamer [26] (*5*, R = *tert*-Butyl) and cyclic heptamer [27] (*7*, R = *tert*-Butyl) under some conditions. In addition, *p-tert*-butyldihomooxacalix[4]arene (*9*) is formed in ponderable quantity in certain cases [23,24,28]. The structures of the cyclic oligomers in the *p-tert*-butyl series have been well established. All are in complete agreement with chemical, spectral, and analytical data; and definitive structure proofs via x-ray crystallography have been provided for compounds *4* (R = *tert*-butyl) [29], *4* (R = 1,1,3,3-tetramethylbutyl) [30], *5* (R = *tert*-Butyl) [31], *5* (R=H) [32], *6* (R = *tert*-Butyl) [33], and *8* (R = *tert*-Butyl) [34].

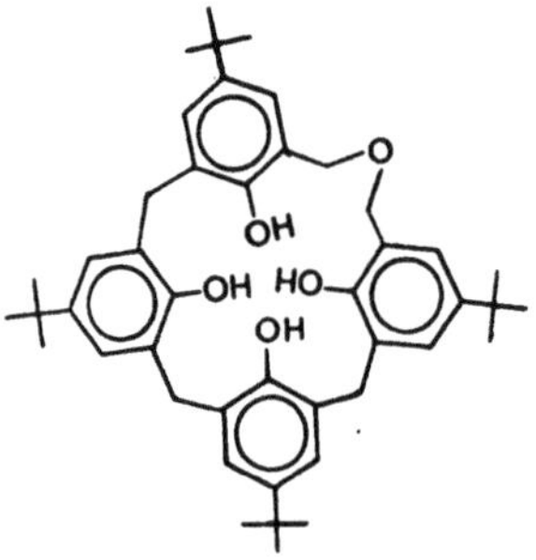

9

Although the structures of the cyclic oligomers from the base-catalyzed condensation of *p-tert*-butylphenol and formaldehyde are now well understood, there is considerable confusion in the older literature concerning these compounds. As early as 1912 Raschig postulated the existence of cyclic compounds in Bakelite products [35], but Baekeland pointed out [36] that "we must not forget that one hypothesis is about as easy to propose as another as long as we are unable to use any of the methods for determining molecular size and molecular constitution". Not until three decades later did evidence begin to accumulate in support of cyclic oligomers as condensation products of phenols and formaldehyde. In 1941 Zinke and Ziegler [15] described a product which they obtained in good yield from a base-catalyzed condensation of *p-tert*-butylphenol and formaldehyde. It was stated to have a melting point above 340 °C and to form an acetate that had a molecular weight of 1725. No structure was suggested for this product, but in retrospect it seems quite certain that what these workers had isolated was *p-tert*-butylcalix[8]arene (*8*, R = *tert*-Butyl). Three years later Zinke and Ziegler again described a *p-tert*-butylphenol/formaldehyde condensation product [16], prepared under somewhat different and more carefully detailed conditions than those previously described, to which they assigned a cyclic *tetrameric* structure (*4*, R = *tert*-Butyl). However, they stated that molecular weight data could not be obtained, because neither the parent compound nor its acetate were sufficiently soluble. Thus, the cyclic tetrameric structure seems to have been based more on intuition than on solid data, cyclic tetrameric structures being "in the air" at the time. For example, Niederl and McCoy [37] claimed to have obtained a *p*-methylcalix[4]arene from an acid-catalyzed condensation of *p*-cresol with 2,6-*bis*-(hydroxymethyl)-4-methylphenol, repeating the work of Koebner [38] who had assigned a linear trimeric structure to this product. But, Koebner's contention was ultimately sustained [39,40], and it is quite certain that a cyclic tetramer is *not* formed under these conditions. Zinke and coworkers continued the investigation of the products of the base-catalyzed reactions of *p*-substituted phenols and formaldehyde and in 1952 published data on the products from the seven phenols cited above. They reported a molecular weight of 873 for the acetate of the product from *p*-octylphenol and formaldehyde, in agreement with a cyclic tetrameric structure. Thus, it was assumed that all of the other *p*-substituted phenol/formaldehyde products were also cyclic tetramers.

The first suggestion that the Zinke products were not pure entities came from Cornforth's experiments [10] in which he isolated mixtures from the condensations of *p*-*tert*-butylphenol and *p*-octylphenol with formaldehyde. The high-melting compounds were designated as HBC and HOC, respectively, and the low-melting compounds as LBC and LOC. Although the molecular weights of all of these compounds and/or their acetates seemed to be in agreement with a cyclic tetrameric structure and although preliminary x-ray crystallographic data seemed also to support this contention, more recent work [20–23,29,30,34] indicates that only the low-melting compounds (LBC and LOC) possess this structure. The high melting compounds (HBC and HOC) are now known to be the cyclic octamers [23,41]. Other workers, including the author of this review, also succumbed to the intuitively appealing and logical assumption that the products of the base-catalyzed condensation of *p*-substituted phenols and formaldehyde must possess cyclic tetrameric structures. Using a condensation procedure devised by chemists of the Petrolite Corporation, Gutsche and coworkers [42] reported the preparation of "cyclic tetramers" from *p*-methyl-, *p*-*tert*-butyl-, *p*-phenyl-, *p*-methoxy-, and *p*-carbomethoxyphenol with formaldehyde. Using a slightly modified version of the Petrolite procedure, Patrick and Egan [43] condensed the same five phenols and also imputed cyclic tetrameric structures to all of the products [44]. Subsequent experiments by Gutsche et al. [23,45,46], however, have shown that in none of these condensations is the cyclic tetramer a major product and that in most instances it is present in such low amounts as to be nonisolable. The Petrolite procedure [47], devised to simulate the factory production of phenol/formaldehyde resins for the manufacture of surfactant compounds, consists of refluxing a *p*-substituted phenol, paraformaldehyde, and a trace of 50% sodium hydroxide in xylene for several hours (the Patrick-Egan modification substitutes potassium *tert*-butoxide for sodium hydroxide and tetralin for xylene). The cooled reaction mixture deposits copious amounts of an insoluble product which, in the case of the *p*-*tert*-butylphenol reaction, is now known to be almost entirely cyclic octamer. Thus, from *p*-*tert*-butylphenol crystalline *p*-*tert*-butylcalix[8]arene (*8*, R = *tert*-Butyl) can be obtained in yields of 60–70%, making it a readily available cyclic oligomer.

In the process of unravelling the intricacies of the condensation of *p*-*tert*-butylphenol and formaldehyde [23], it was discovered that if a stoichiometric amount of base is used in the condensation instead of the catalytic amount employed in the original Petrolite procedure the major product is the cyclic hexamer, *p*-*tert*-butylcalix[6]arene (*6*, R = *tert*-Butyl). Yields as high as 70–75% of pure, crystalline material can be obtained, thus making this another abundantly available cyclic oligomer. Ironically, the cyclic tetramer is the even-numbered cyclic oligomer produced in lowest yield. Employing the Zinke procedure as modified by Cornforth by the substitution of Dowtherm (a eutectic of biphenyl and diphenyl ether) for linseed oil in the final step, one can obtain *p-tert*-butylcalix[4]arene in capriciously varying yields ranging from almost nothing to as high as 45%. Considerable effort has been expended in an attempt to understand the details of this reaction, but definitive results have yet to be obtained. One of the critical steps in the Zinke-Cornforth procedure is the last one in which the solid resinous material is powdered and heated (i.e. in linseed oil or Dowtherm). With regard to this step Zinke states [19] that "we believe we have isolated such cyclized compounds by heating resoles which had been condensed as far as the ether stage and which had not been washed free from alkali". Experiments in our laboratories have shown that acid washing the finely powdered resin fails to remove all of the base (sodium content of acid-washed resin is 1.3%). Only by dissolving the resin in an organic solvent, washing the solvent with acid followed by water, and evaporating the solvent can a base-free resin be obtained. The base-free material fails to yield cyclic oligomers when heated in Dowtherm, but upon the addition of a small amount of base (0.15 equivalent, based on the starting phenol, may be the optimum quantity [48]) cyclic tetramer is formed in 25–35% yield [48,49].

The odd-numbered calixarenes are more difficult to obtain in quantity than the even-numbered calixarenes. Employing the Patrick and Egan modification of the Petrolite procedure and changing the heating sequence (55 °C for 6 hours followed by 150 °C for 6 hours) Ninagawa and Matsuda [26] obtained a mixture from which they isolated 23% cyclic tetramer, 5% cyclic pentamer, and 11% cyclic octamer. Employing the Petrolite procedure but with dioxane as the solvent and a 30 hour heating period, Nakamoto and Ishida [27] obtained a mixture containing cyclic hexamer, heptamer, and octamer from which they isolated 6% of the heptamer.

Little is known about the overall mechanism of cyclic oligomer formation. although the mechanism of the initial stages of the sequence seems fairly clear. The first chemical event is the reaction of formaldehyde (formed in the Petrolite procedures by depolymerization of paraformaldehyde) with phenol to form 2-hydroxymethyl- and 2,6-*bis*(hydroxymethyl)phenols in a base-catalyzed process, as shown in Fig. 3. Such compounds were characterized many years ago [50], obtained from the action of aqueous formaldehyde on phenol in basic solution at room temperature. Subsequent condensation between the hydroxymethylphenols and the starting phenol occurs to form linear dimers, trimers, tetramers, etc. via a pathway that might involve *o*-quinonemethide intermediates which react with phenolate ions in a Michael-like reaction, as portrayed in Fig. 4. The condensation of hydroxymethyl-

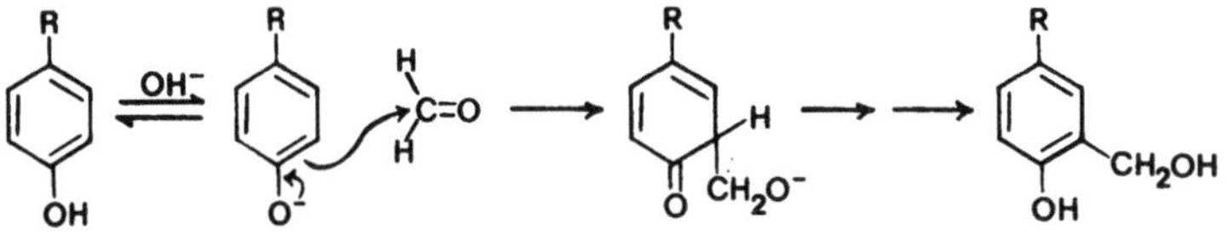

Fig. 3. Base-catalyzed hydroxymethylation of phenols

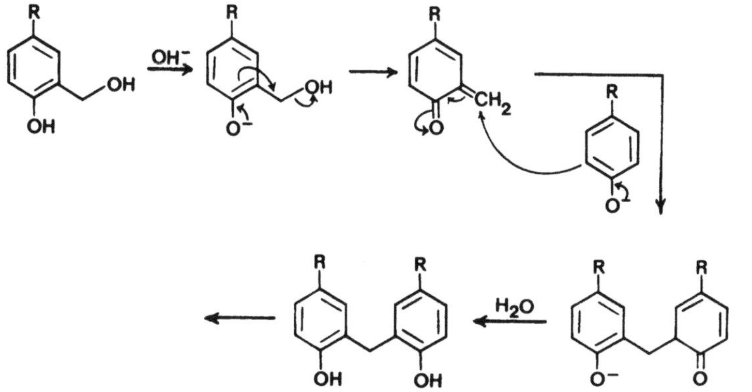

Fig. 4. Base-catalyzed formation of linear oligomers from phenols and formaldehyde

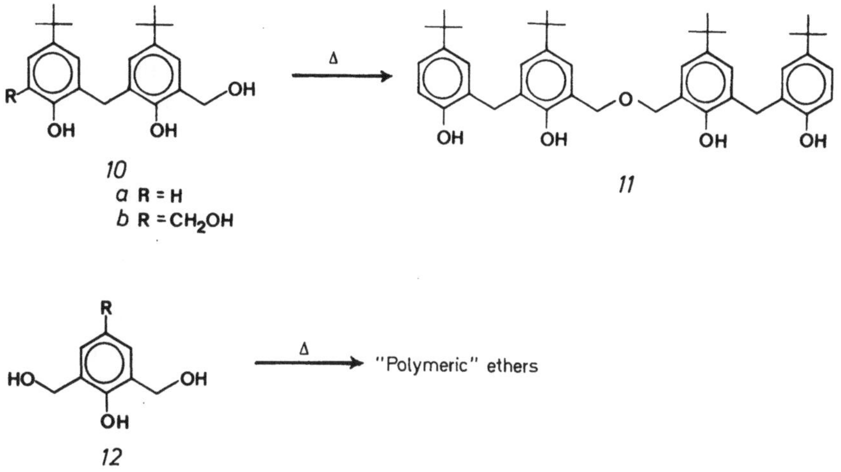

Fig. 5. Formation of dibenzyl ethers of 2-hydroxymethylphenols

phenols to form oligomers also has been shown to occur under relatively mild conditions. The possibility of quinonemethide intermediates was suggested as far back as 1912[51] and has been promoted by Hutzsch[52], v. Euler[53], and others. However, because the formation of *o*-quinonemethides from compounds such as *o*-(methoxymethyl)phenol requires quite high temperatures (500–600 °C)[54] doubt has been cast on the validity of this premise[55]. On the other hand, it is known that oxy-Cope rearrangements occur far more readily with anions than with the corresponding neutral compounds[56], so conversion of 2-hydroxymethylphenolates to *o*-quinonemethides may, in fact, occur under the basic conditions prevailing in the reactions under discussion. Intermolecular dehydration of 2-hydroxymethylphenols to form dibenzyl ethers also occurs under the conditions of the Zinke-Cornforth and Petrolite condensation reactions, as illustrated in Fig. 5. For example, *10a* yields the ether *11*[57], and *12* (R = CH_3 and R = *tert*-Butyl) yields polymeric ethers[58 59] upon simple heating. Thus, the mixtures from which the calixarenes emerge contain linear oligomers of various lengths in which the *o,o'*-bridges are CH_2 as well as CH_2OCH_2 groups.

The events that occur in the terminal phases of the sequence leading to the cyclic oligomers remain a mystery. It is certain that the linear oligomers lose water and formaldehyde in the process of being converted to the cyclic oligomers [60], but the immediate precursors of the cyclic oligomers are not known. The preferential formation of the even-numbered cyclic oligomers in the *p-tert*-butyl series might suggest a common precursor such as *10a* or *10b* which, by dimerization, trimerization, and tetramerization could yield *4*, *6*, and *8* (R = *tert*-butyl). The detection of cyclic pentamer and cyclic heptamer in the reaction mixtures, however, casts doubt on this pathway. Also uncertain is the nature of the driving forces that promote cyclization in high yields, particularly in the case of the cyclic hexamers and octamers; why are the larger sized cyclic oligomers formed with greater ease than the cyclic tetramer, which should be strongly favored on entropic grounds? Our current hypothesis invokes a combination of intramolecular hydrogen bonding and cation template phenomena. The cyclic tetramer is strongly intramolecularly hydrogen bonded, as indicated by the concentration-independent OH stretching absorptions at 3160 cm^{-1} in the infrared (see Sect. 4.4). Inspection of space-filling molecular models (CPK models) shows that the four OH groups of the cyclic tetramer are forced into close proximity (in the "cone" conformation; see Sect. 5.1). Surprising, however, is the fact that the more flexible cyclic hexamer and octamer show similar IR behavior, having OH stretching absorptions at 3150 cm^{-1} and 3230 cm^{-1}, respectively and indicating very strong intramolecular hydrogen bonding in these systems as well. Even more surprising is the fact that the linear tetramer, pentamer, and hexamer have OH stretching absorptions at 3200 cm^{-1}, again indicative of strong intramolecular hydrogen bonding [61]. Hydrogen bonding in the linear oligomers can either be intermolecular, giving rise to pseudocyclic arrays which we have designated as "hemicalixarenes" or intramolecular, giving rise to pseudocyclic arrays which we have designated as "pseudocalixarenes" [62], as illustrated in Fig. 6. Pseudocalixarene formation might constitute a major factor in determining the course of the cyclization process.

It is reported [23] that the yield of cyclic hexamer, which is the major product when a stoichiometric amount of base is used, is slightly better with RbOH than with CsOH, KOH, or NaOH; LiOH is an ineffective catalyst. This has been interpreted as suggesting that a template effect may play a part in the cyclization

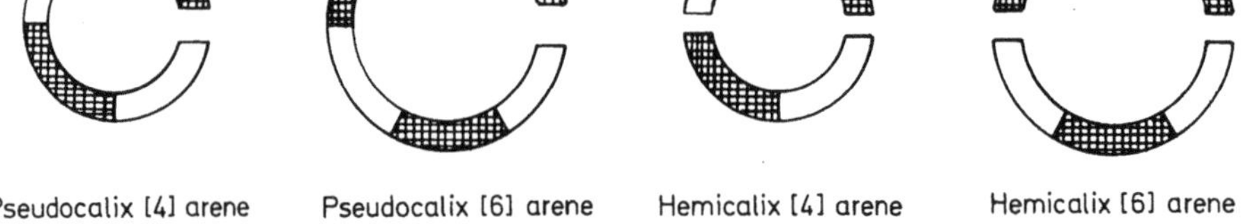

Fig. 6. Hemicalixarenes and pseudocalixarenes

process. The work of Izatt and coworkers [63] has shown that the calixarenes do, indeed, have ionophoric capacity (see Sect. 7.3) for NaOH, KOH, RbOH, and CsOH but nor for LiOH. The details of the role of cations in the cyclization process, however, remain to be clarified.

An observation that promises to have significant mechanistic implications concerns the interconvertibility of the cyclic oligomers. Contrary to an earlier report that they are non-interconvertible under the conditions of the reaction [23], it has recently been demonstrated [49] that when *p-tert*-butylcalix[8]arene and *p-tert*-butylcalix[6]arene are heated in boiling diphenyl ether in the presence of a small amount of potassium *tert*-butoxide a 20–35% yield of *p-tert*-butylcalix[4]arene is obtained.

3.1.2 Acid-Catalyzed Condensation of Resorcinols and Aldehydes

The acid-catalyzed reactions of *p*-substituted phenols and formaldehyde yield mixtures of linear oligomers which, under certain conditions [64], include compounds containing as many as 25 or more monomeric units. There is no evidence that any cyclic oligomers are present in these mixtures [65]. Resorcinol, on the other hand, has long been known to react with aldehydes (other than formaldehyde) to yield well defined materials. In 1883 [66] Michael isolated two crystalline compounds from the acid-catalyzed reaction between resorcinol and benzaldehyde, and he assigned a cyclic dimeric structure to one of them. In 1940 this assignment was reformulated by Niederl and Vogel [67] to a tetrameric structure, *i.e.* a calix[4]arene containing eight *extra*-annular hydroxyl groups [12], as illustrated in Fig. 7 for compound *13* (R′ = H). Mass spectrometric support was supplied by Erdtman and coworkers [68] on the octamethyl ether (*13*, R = R′ = CH_3), and x-ray crystallographic determinations [69,70] conclusively established the structures.

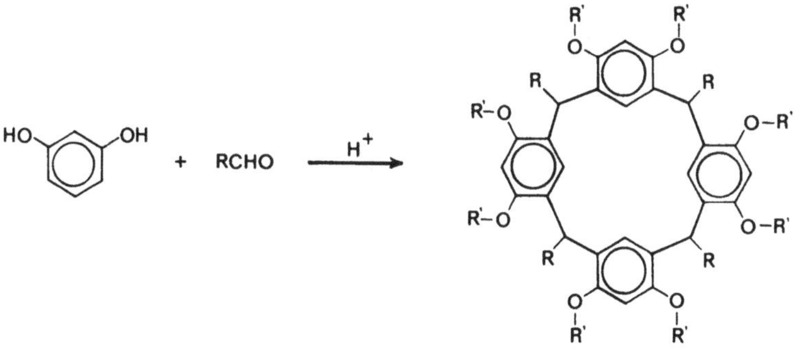

13

Fig. 7. One-step synthesis of octahydroxycalix[4]arenes

The mechanism of the reaction has been studied in some detail by Högberg [12,71,72]. In contrast to the base-catalyzed oligomerization, the acid catalyzed process involves electrophilic aromatic substitutions by cations, as outlined in Fig. 8. Although formaldehyde does not react with resorcinol to produce cyclic oligomers, other aldehydes such as acetaldehyde and benzaldehyde give excellent yields of

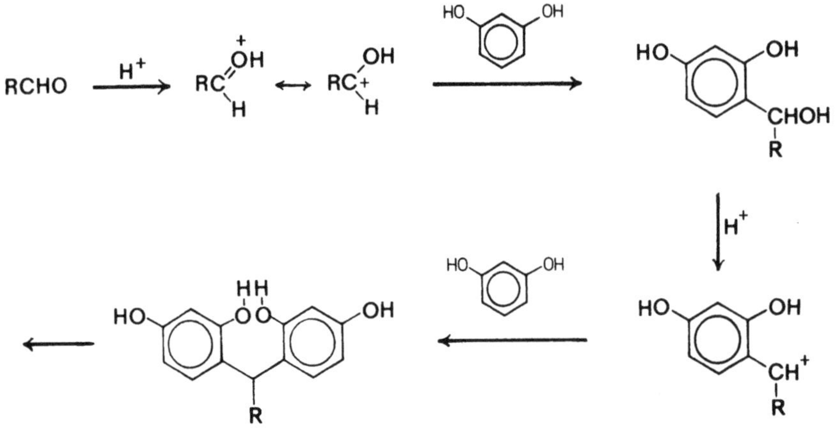

Fig. 8. Acid-catalyzed formation of resorcinol/aldehyde oligomers

cyclic tetramer *13*. The problem with formaldehyde arises from its great reactivity which leads to substitution at the 2-position as well as the 4- and 6-positions, resulting in the formation of cross-linked polymers. Other aldehydes, more bulky than formaldehyde, are less likely to react at the hindered 2-position. Of the four diastereomeric possibilities for *13*, assuming conformational mobility (see Sect. 5.1), only two comprise the bulk of the product, *viz. cis-cis-cis* and *cis-trans-cis*, presumably because of non-bonded interactions in the precursors leading to the final products. Högberg has demonstrated that the *cis-trans-cis* isomer is the kinetic product in the case of the resorcinol/benzaldehyde condensation but that it can be converted, *in situ*, to the *cis-cis-cis* product which is the more insoluble and which separates from solution. By taking advantage of this circumstance, a greater than 80% yield of the *cis-cis-cis* isomer can be obtained [72]. In the case of the product from 2-methylresorcinol and benzaldehyde [12] it is the *cis-trans-cis* isomer that is the more insoluble and, therefore, the one that can be obtained in high yield. The oligomerization in these cases is a reversible process [73], and insolubility is a driving force for the formation of a single isomer and, perhaps, for the cyclization itself. It is uncertain whether intramolecular hydrogen bonding in the acyclic precursor(s) plays a part comparable to that imputed to the base-catalyzed process involving *p*-substituted phenols and formaldehyde. The IR stretching frequencies reported [71] for the isomers of *13*, which are very broad (3700–2500 cm^{-1}), do not provide any clear-cut information on this point, and the IR data on the corresponding linear oligomers are not available.

3.1.3 Acid-Catalyzed Condensations of Alkylbenzenes and Formaldehyde

Calix[4]arenes *14a* and *14b* have been prepared from mesitylene and from 1,2,3,5-tetramethylbenzene by condensation with formaldehyde in the presence of acetic acid [74]. Calixarene *14a* has also been prepared by a Friedel-Crafts reaction with chloromethylmesitylene [75].

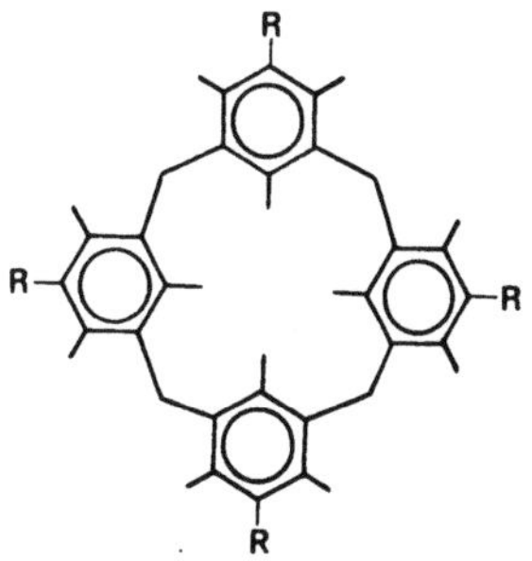

14a (R = H)

14b ($R = CH_3$)

3.1.4 Acid-Catalyzed Condensations of Heterocyclic Compounds and Aldehydes

Although not formally classified as calixarenes, compounds closely related in architecture to the calix[4]arenes have been prepared by the condensation of furans, thiophenes, and pyrroles with aldehydes and ketones. Furan undergoes acid-catalyzed condensation with aldehydes and ketones to give *15* [76–79], as shown in Fig. 9.

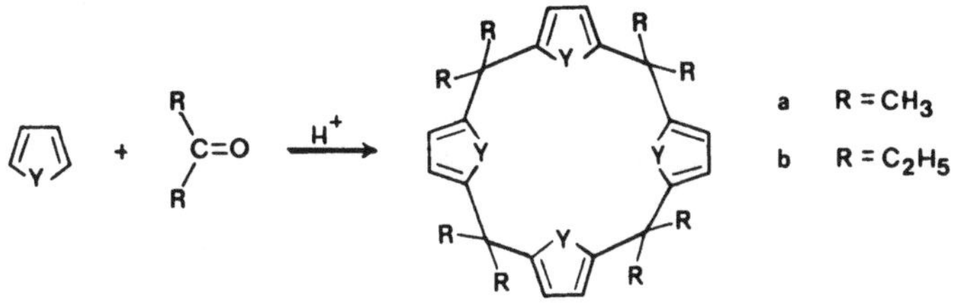

15 Y = O

16 Y = S

17 Y = N

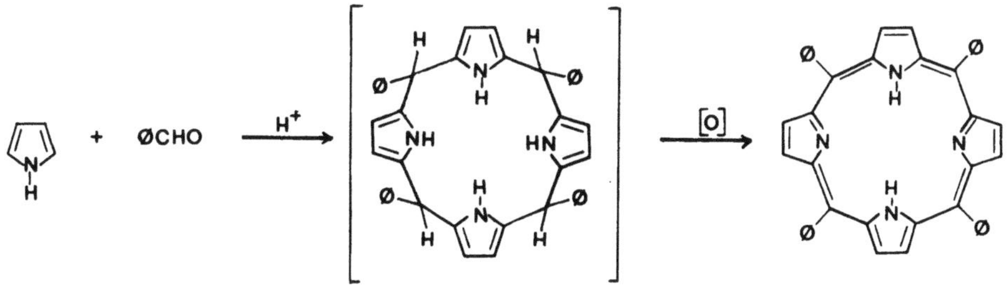

18

Fig. 9. Acid-catalyzed formation of cyclic tetramers from furan, thiophene, and pyrrole with aldehydes and ketones

The increase in yield of *15a* from *ca* 20% to over 40% in the presence of lithium perchlorate [77] has led to the suggestion of a template effect. However, recent experiments [78] indicate that the higher yields correlate not with the metal ion but with the acidity of the reaction medium. In similar fashion, thiophene and pyrrole undergo acid-catalyzed condensations with acetone to yield *16* [80] and *17* [81]. Benzaldehyde also condenses with pyrrole [82], but the initially formed compound loses six hydrogens to form the planar tetraphenylporphin *18*.

3.2 Related Condensations Involving Formaldehyde

Formaldehyde is an exceedingly reactive molecule and condenses with a wide variety of compounds, some of which have been illustrated by the examples in the previous parts of this Section. Other examples of formaldehyde condensations leading to macrocyclic compounds include veratrole (leading to cyclotriveratrylene [83]), glycoluril (leading to curcurbituril [84]), 2,5-dimethylthiophene (leading to a cyclic trimer [85]), 2-arylimidazole (leading to a cyclic trimer and a cyclic tetramer [86]), and N-methylindole (leading to a cyclic trimer [87]).

3.3 Stepwise Synthesis of Calixarenes

3.3.1 Hayes-Hunter-Kämmerer Synthesis

Noting Zinke's claim to have isolated cyclic tetramers from the condensation of *p*-substituted phenols and formaldehyde, Hayes and Hunter in 1956 sought to

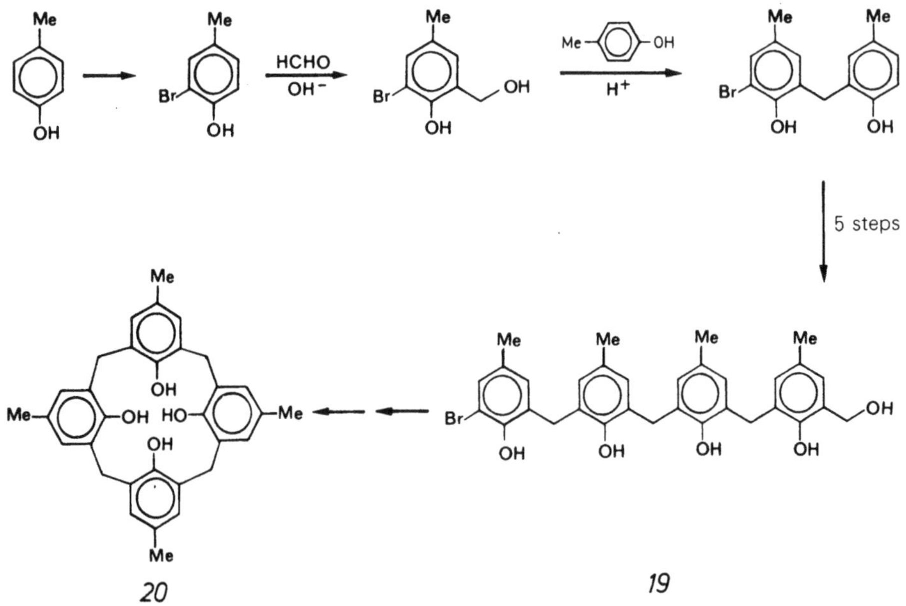

Scheme 1. 10-Step synthesis of *p*-methylcalix[4]arene (Hayes and Hunter method).

provide additional evidence for such structures by what they termed a "rational" synthesis [9]. Starting with *p*-cresol, they protected one of the *o*-positions by bromination and then sequentially added methylene groups (by base-induced hydroxymethylation) and aryl groups (by acid-catalyzed arylation). The *o*-bromo-*o'*-hydroxymethyl linear tetramer (*19*) thus obtained was debrominated and then cyclized to *20*, as illustrated in Scheme 1. Although no comparison of the Hayes and Hunter product with the Zinke product was reported, this synthesis seems to have been generally accepted as implicit proof for the Zinke tetrameric structure [88]. More recently, Kämmerer and coworkers [20,21,89–94] have improved and expanded the Hayes and Hunter synthesis, demonstrating its potential by the preparation of a series of methyl- and *tert*-butyl-substituted calixarenes, including the cyclic tetramers, pentamers, hexamers, and heptamers shown in Fig. 10.

The Hayes and Hunter stepwise synthesis has certain drawbacks; it is long and

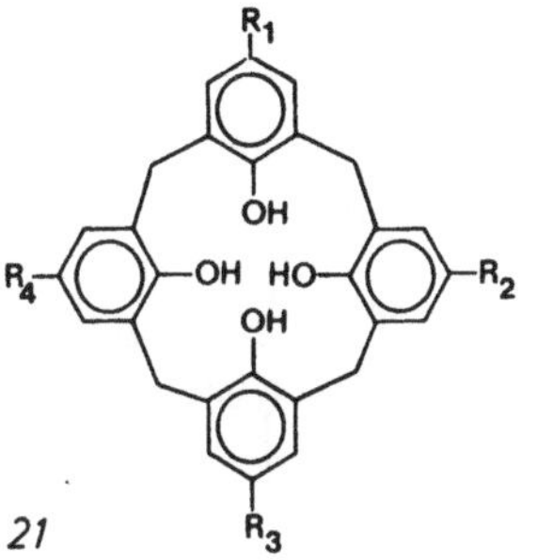

21

21	*a*	*b*	*c*	*d*	*e*	*f*	*g*
R_1	H	Me	Me	Me	Me	Me	t-Bu
R_2	H	Me	Me	Me	t-Bu	t-Bu	t-Bu
R_3	Me	Me	Me	t-Bu	Me	t-Bu	t-Bu
R_4	Me	Me	t-Bu	t-Bu	t-Bu	t-Bu	t-Bu
R_5							
R_6							
Ref.	90)	20)	89)	89,90)	89)	21,89)	21,89,90)

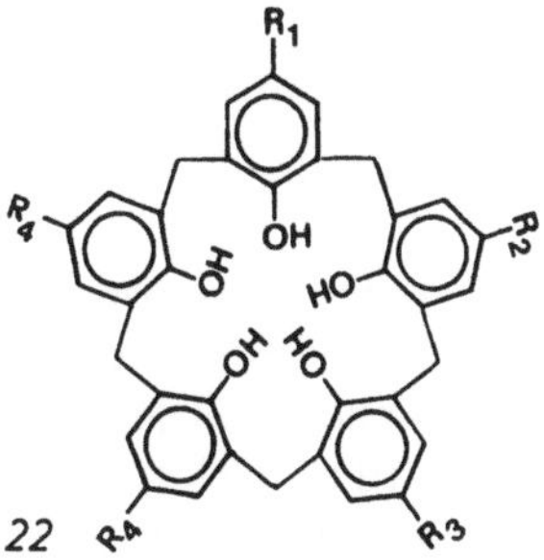

22

22	*a*	*b*	*c*
R_1	Me	t-Bu	t-Bu
R_2	Me	Me	t-Bu
R_3	Me	Me	t-Bu
R_4	Me	Me	Me
R_5			
R_6			
Ref.	21,93)	93)	93)

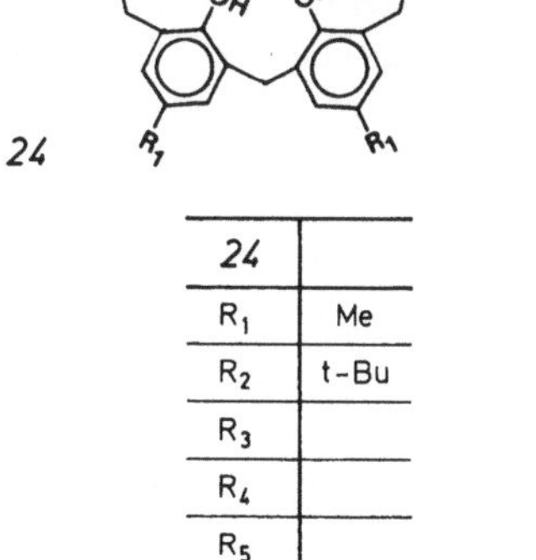

24

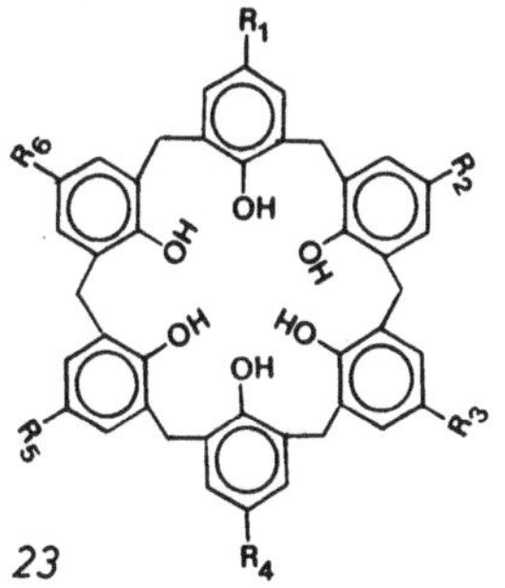

23

23	*a*	*b*	*c*
R_1	Me	t-Bu	t-Bu
R_2	Me	Me	Me
R_3	Me	Me	Me
R_4	Me	Me	t-Bu
R_5	Me	Me	Me
R_6	Me	Me	Me
Ref.	92)	92)	92)

24	
R_1	Me
R_2	t-Bu
R_3	
R_4	
R_5	
R_6	
Ref.	91)

Fig. 10. Calixarenes synthesized by Kämmerer et al. via the Hayes and Hunter stepwise method

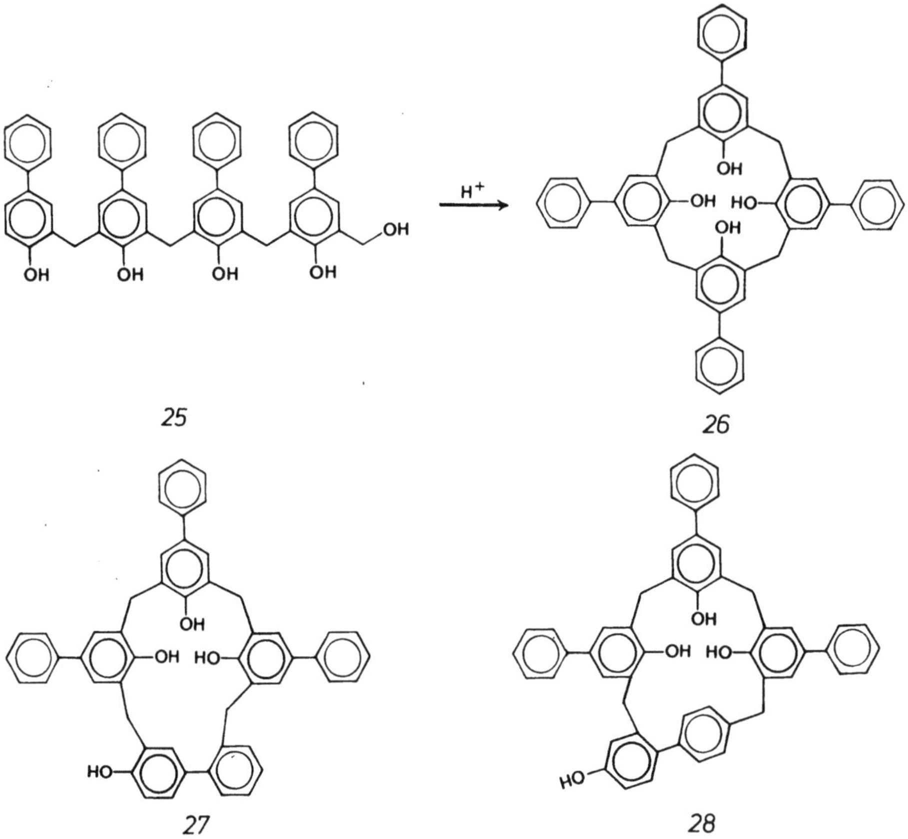

Scheme 2. Stepwise synthesis of *p*-phenylcalix[4]arene by the Hayes and Hunter method.

it affords the final product in poor to modest overall yield. For example, the conversion of *p-tert*-butylphenol to *p-tert*-butylcalix[4]arene [23] proceeds in about 11% overall yield, while the conversion of *p*-phenylphenol to *p*-phenylcalix[4]arene proceeds in only 0.5% overall yield [95]. The latter case is complicated by the formation of three compounds in the cyclization reaction, as shown in Scheme 2. One is the desired product (*26*), and the other two are postulated to be the isomeric compounds *27* and *28*, formed as the result of cyclization into the other reactive sites of the terminal *p*-phenylphenyl moiety. Thus, *p*-phenylcalix[4]arene, a compound of considerable interest with respect to the formation of molecular complexes, is not readily accessible via this route.

3.3.2 Böhmer, Chhim, and Kämmerer Synthesis

Recognizing the deficiencies in the Hayes and Hunter synthesis, Böhmer, Chhim, and Kämmerer [96] have explored a more convergent approach which retains much of the flexibility of the sequential approach. It involves the condensation of a linear trimer (*29*) with a 2,6-*bis*halomethylphenol (*30*), as illustrated in Scheme 3. Although short, this method suffers from quite low yields in the cyclization step, ranging from

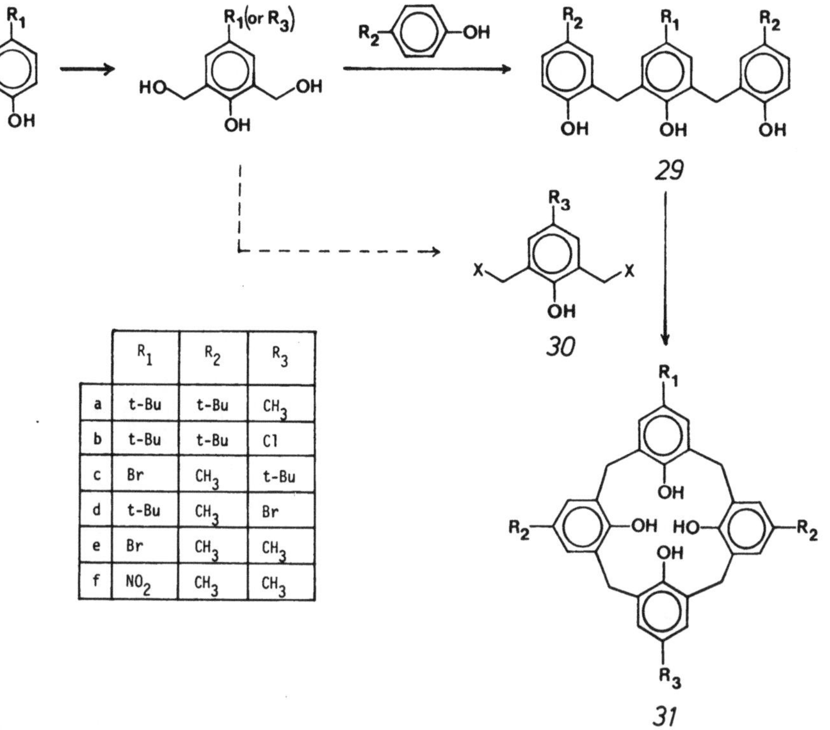

	R_1	R_2	R_3
a	t-Bu	t-Bu	CH_3
b	t-Bu	t-Bu	Cl
c	Br	CH_3	t-Bu
d	t-Bu	CH_3	Br
e	Br	CH_3	CH_3
f	NO_2	CH_3	CH_3

Scheme 3. Convergent, stepwise synthesis of calixarene[4]arenes (Böhmer, Chhim, and Kämmerer method).

10–20% in the best cases down to 2–7% in the more interesting cases in which a mixture of alkyl, bromo, and nitro functions are incorporated as R groups.

3.3.3 Moshfegh, Hakimelahi et al. Synthesis

As part of an extensive program dealing with analogues of phloroglucides, Moshfegh, Hakimelahi and coworkers [97–99] have investigated the synthesis of a variety of *p*-halocalix[4]arenes using a procedure similar to that of Böhmer et al. When the dimer *32* and the *bis*(hydroxymethyl) dimer *33* are used as starting materials calixarene *34* is stated to be produced in 65% yield (however see Ref. [124]), accompanied by 10% of the diooxa compound *35*. In similar fashion the condensation of the dimer *32* (X = Cl) with 4-halo-2,6-*bis*(hydroxymethyl)phenol (*36*) is stated to

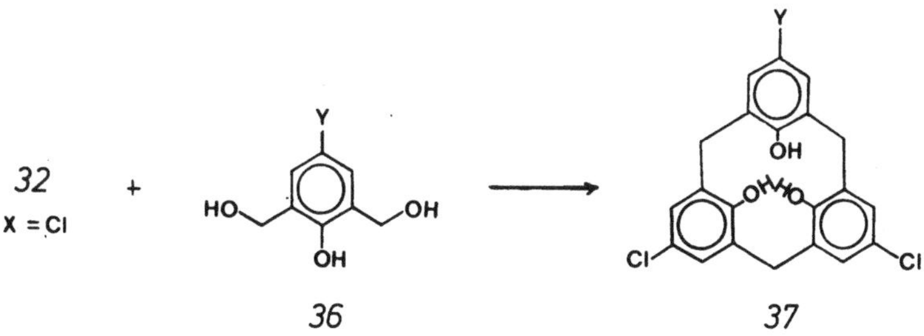

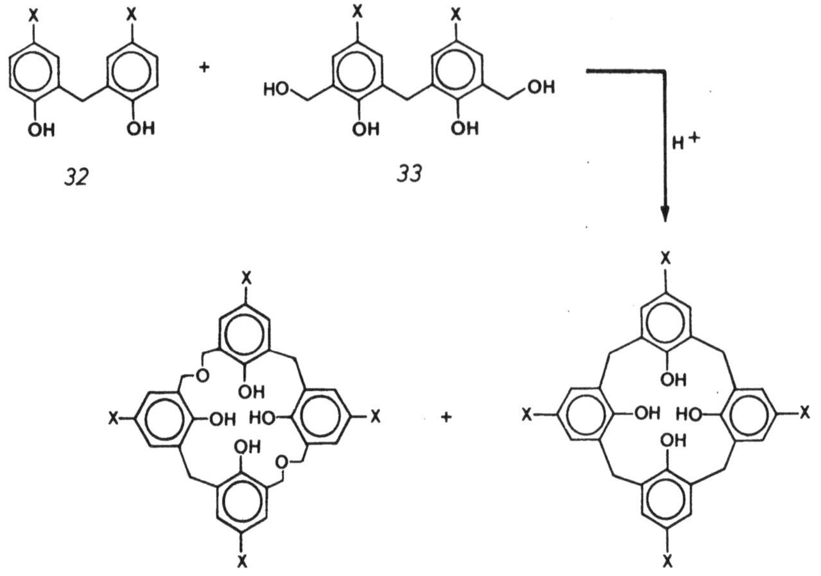

32 33 35 34

Scheme 4. Convergent, step-wise synthesis of calix[4]arenes (Moshfegh, Hakimelahi et al. method).

afford a 69–90% yield of *p*-halocalix[3]arenes (*37*). The striking contrast in yield between the Böhmer, Chhim, and Kämmerer method and the Moshfegh, Hakimelahi et al. method for the preparation of calix[4]arenes indicates that a careful investigation should be made to determine the true potentialities of this general approach. Also, since space filling models of the calix[3]arene ring system indicate considerably more non bonded strain than in the higher membered calixarenes, it is hoped that additional and more complete details in support of this structure will be forthcoming.

3.3.4 No and Gutsche Synthesis

Another convergent synthesis has been devised by No and Gutsche [100] which retains some of the functional group flexibility of the Hayes and Hunter method and which gives sufficiently high yields to make calixarenes available in quantities large enough for catalysis studies. In the 4-step sequence shown in Scheme 5 a *p*-substituted phenol (*e.g.* *p*-phenylphenol) is treated with formaldehyde under controlled conditions to produce the *bis*(hydroxymethyl) dimer *38*; the dimer is condensed with two equivalents of a *p*-substituted phenol (*e.g.* *p*-*tert*-butylphenol) to yield the linear tetramer *39* (*e.g.* R = *tert*-Butyl); and, the tetramer is monohydroxymethylated to yield *40* (*e.g.* R = *tert*-Butyl) and then cyclized to yield calix[4]arene (*e.g.* R = *tert*-Butyl). Although the overall yield is only ca 10%, the starting materials are cheap and the work-up and purification procedures are generally simple and straightforward. If improvements in the selective hydroxymethylation step can be realized the synthesis would excel the Zinke "one-flask" method not only with respect to flexibility but with respect to yield as well.

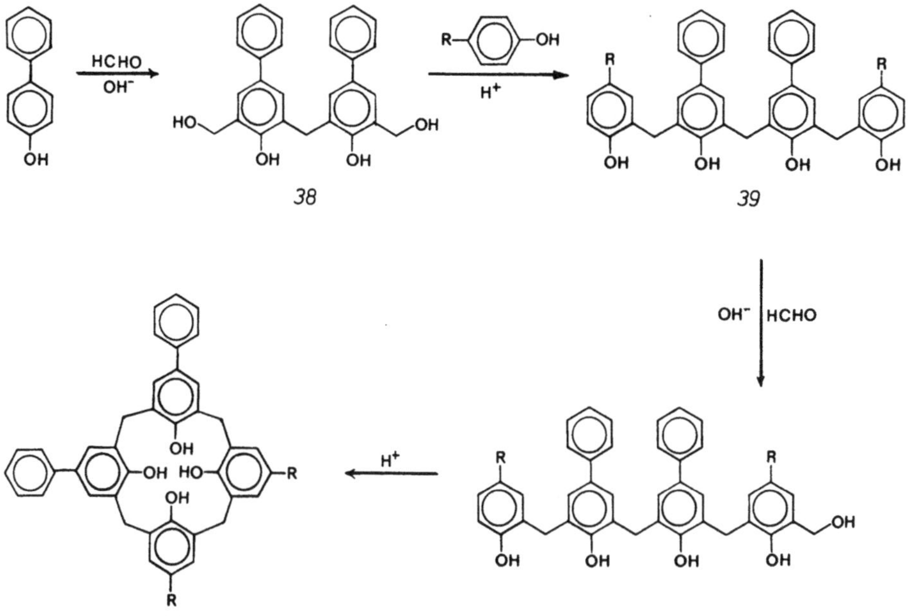

38
39
41
40

Scheme 5. Convergent, step-wise synthesis of calix[4]arenes (No and Gutsche method).

3.4 Oxacalixarenes

The Petrolite method for calixarene synthesis also leads to dihomooxacalix[4]arenes, as represented by the *p-tert*-butyl compound *9*. Proton-NMR spectra and TLC data indicate that the mixture obtained from *p-tert*-butylphenol and paraformaldehyde contains appreciable amounts of *9*, but it is difficult to obtain a pure sample of *9* from this mixture [23,24,28]. Compound *9* is much more readily accessible by the thermally-induced dehydration of the *bis*(hydroxymethyl)tetramer *46* which can be readily prepared by the convergent synthesis outlined in Scheme 5 followed by *bis* hydroxymethylation [62], as shown in Scheme 6. Other oxacalixarenes can also be obtained by thermally-induced dehydration. For example, 2,6-*bis*(hydroxymethyl)-phenols (*42*) yield hexahomotrioxacalix[3]arenes (*43*) [28,62,101]; *bis*(*hyd*roxymethyl) dimers *44a* yield tetrahomodioxacalix[4]arenes *45a* [62,97,102]; and *bis*(hydroxymethyl) dimers *44b* yield octahomotetraoxacalix[4]arenes *45b* [103]; as illustrated in Scheme 6. The ease with which these dehydrations occur has been attributed [62] to intermolecular and intramolecular hydrogen bonding forces which establish the reacting systems in cyclic arrays prior to cyclization (see Fig. 6).

3.5 Calixarene Esters and Ethers

The phenolic OH groups of the calixarenes are readily converted to ester and ether moieties. The acetates of many of the calixarenes have been prepared, the earliest examples being those reported by Zinke and Ziegler [15,16]. The acetates are frequently, but not invariably, lower melting than the parent calixarene and are more soluble in organic solvents. Advantage can be taken of the improved solubility in the purification of calixarene mixtures produced in the "one-flask" procedures. For example, the product obtained from *p*-phenylphenol is not amenable to purification by crystallization, fractional extraction, or column chromatography. Conversion of the crude mixture to the acetate gives a more tractible material from which pure compounds have been isolated [46] by flash chromatography [104]; the parent calixarenes can then be obtained by removal of the ester groups by means of

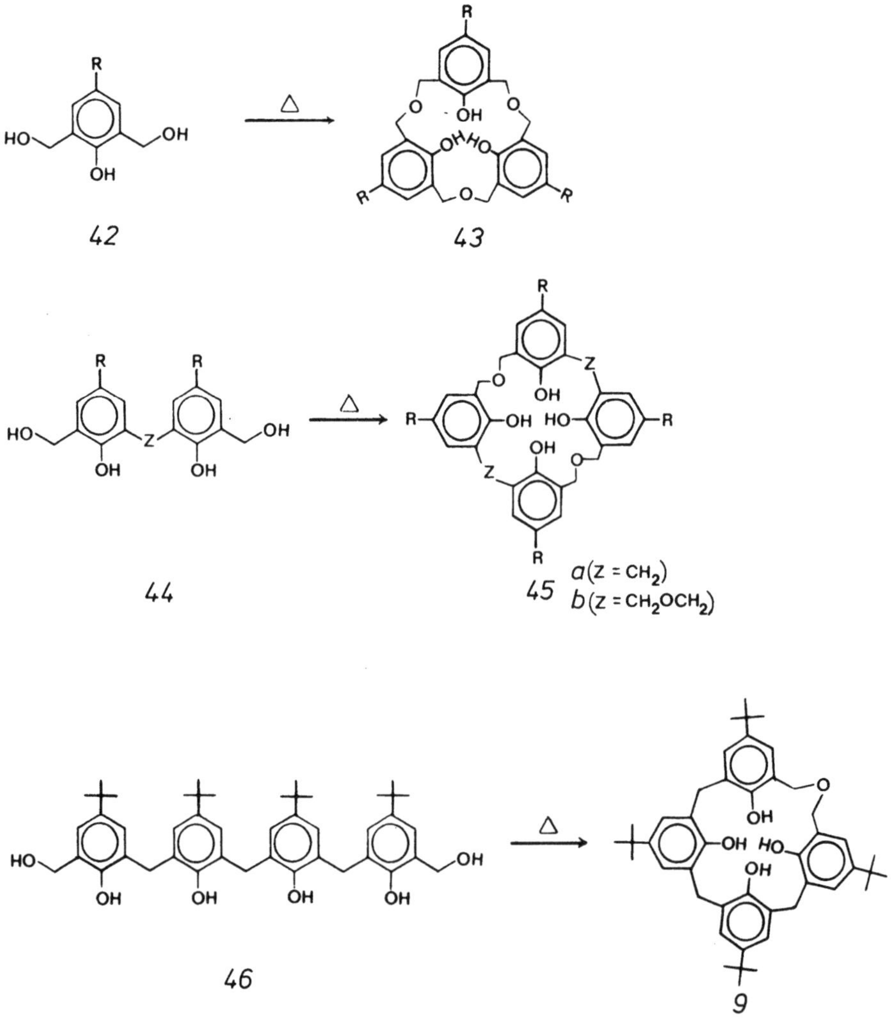

Scheme 6. Synthesis of oxacalixarenes by intra- and intermolecular dehydration.

ethylene diamine in DMF solution [105]. When an excess of acetylating agent is used the completely acetylated product is generally produced. However, instances of incompletely acetylated calixarenes have been reported. For example, two acetates have been isolated from *p-tert*-butylcalix[4]arene that are not conformational isomers (see Section 5.2) [23]; the one melting at 383–386° is assigned the tetraacetate structure [106, 107], and the one melting at 247–250 °C the triacetate structure [107]. Although the acetates are the most commonly prepared ester of calixarenes, others such as the benzoates and *p*-toluenesulfonates can be easily made [48, 65]. The mono- and di-camphorsulfonyl esters of *p-tert*-butylcalix[8]arene have been reported and the effects on the circular dichronic characteristics of the chiral camphorsulfonyl moiety noted [25].

A convenient method for preparing the completely alkylated ethers of the calixarenes involves treatment of the calixarene in THF-DMF solution with an alkyl halide in the presence of sodium hydride. Methyl, ethyl, allyl, and benzyl ethers have all been prepared in this fashion in excellent yields [107]. Various $(CH_2CH_2O)_nR$ ethers have been prepared via the action of the tosylate of the alkylating agent in the presence of potassium *tert*-butoxide [106, 108]. Under different conditions, partially alkylated calixarenes have been isolated. For example, treatment of *p-tert*-butylcalix[4]-arene with dimethyl sulfate in the presence of $BaO—Ba(OH)_2$ in DMF yields the trimethyl ether [107], and treatment with benzyl tosylate yields the dibenzyl ether [107]. Treatment with ethereal diazomethane yields a product possessing a ^1H nmr spectrum compatible with a dimethyl ether structure, but an x-ray crystallographic determination carried out by Professor G. G. Stanley of Washington University indicates the structure to be a monomethyl ether. To explain these data it is postulated that the product actually is a mixture in which the major component is the dimethyl ether but that the single crystal selected for x-ray analysis was the monomethyl ether.

2,4-Dinitrophenyl ethers of *p*-tert-butylcalix[8]arene have been prepared by heating a pyridine solution of the calixarene with 2,4-dinitrochlorobenzene [25]. Depending on the ratio of calixarene to arylating agent, the product contains one, two, or six 2,4-dinitrophenyl moieties. Although an excess of arylating agent fails to yield an octa-substituted compound, the remaining OH groups can be converted to acetoxy groups by treatment with acetyl chloride. It is of historical interest that these experiments gave one of the early clues that the materials previously thought to be cyclic tetramers are, in fact, cyclic octamers.

Still another calixarene derivative of interest and utility is the trimethylsilyl ether. The hexa-trimethylsilyl calix[6]arenes and octa-trimethylsilyl calix[8]arenes can be prepared [23] by using standard trimethylsilylating agents such as hexamethyldisilazene and chlorotrimethylsilane. The tetra-trimethylsilyl ethers of the calix[4]arenes, however, do not form under these conditions and require the use of the very reactive N,O-*bis*(trimethylsilyl)acetamide [109].

The formation of calixarene esters and ethers can be complicated by the problem of incomplete derivatization, as noted above, and also by the fact that conformational "fixing" generally occurs in the case of the calix[4]arenes. The consequences of the latter are discussed in Section 5.2.

4 Physical and Spectral Properties of Calixarenes

4.1 The Shape of the Calixarenes

Perceiving a similarity between the shape of a Greek vase known as a Calix Crater and the shape of the cyclic tetramer, as illustrated in Fig. 11., we assigned the name "calixarene" [3,4]. If the calixarenes assume the shape designated as the "cone" conformation (see Sect. 5.1) they are seen to have cavities whose dimensions increase as the number of arene moieties in the macrocyclic array increases, as illustrated in Fig. 12. Whether the shapes of these cavities are time-invariant depends on the flexibility of the calixarene (see Sect. 5.2); whether the "open" conformation exists depends on intramolecular hydrogen bond interactions. There is evidence (see Sect. 5.1), for example, that the calix[6]arenes and calix[8]arenes exist in solution in nonpolar solvents in transannularly "pinched" conformations, as illustrated in Fig. 13. x-Ray crystallographic determinations have established that in the solid state the calix[4]-arenes exist in the "cone" conformation [29,30], the calix[5]arenes in the "cone" conformation [31,32], and the derivatives of the calix[6]arenes and calix[8]arenes in the "alternate" rather than "cone" conformation [33,34].

Fig. 11. *p*-Phenylcalix[4]arene (left) and Calix Crater (right)

4.2 Melting Points of the Calixarenes

A characteristic feature of the calixarenes is their unusually high melting points, almost invariably higher than those of their acyclic counterparts. With the exception of the *p*-halocalix[4]arenes and the *p*-halocalix[3]arenes reported by Moshfegh, Hakimelahi and coworkers [97,99,124], all of the calixarenes prepared to date have melt-

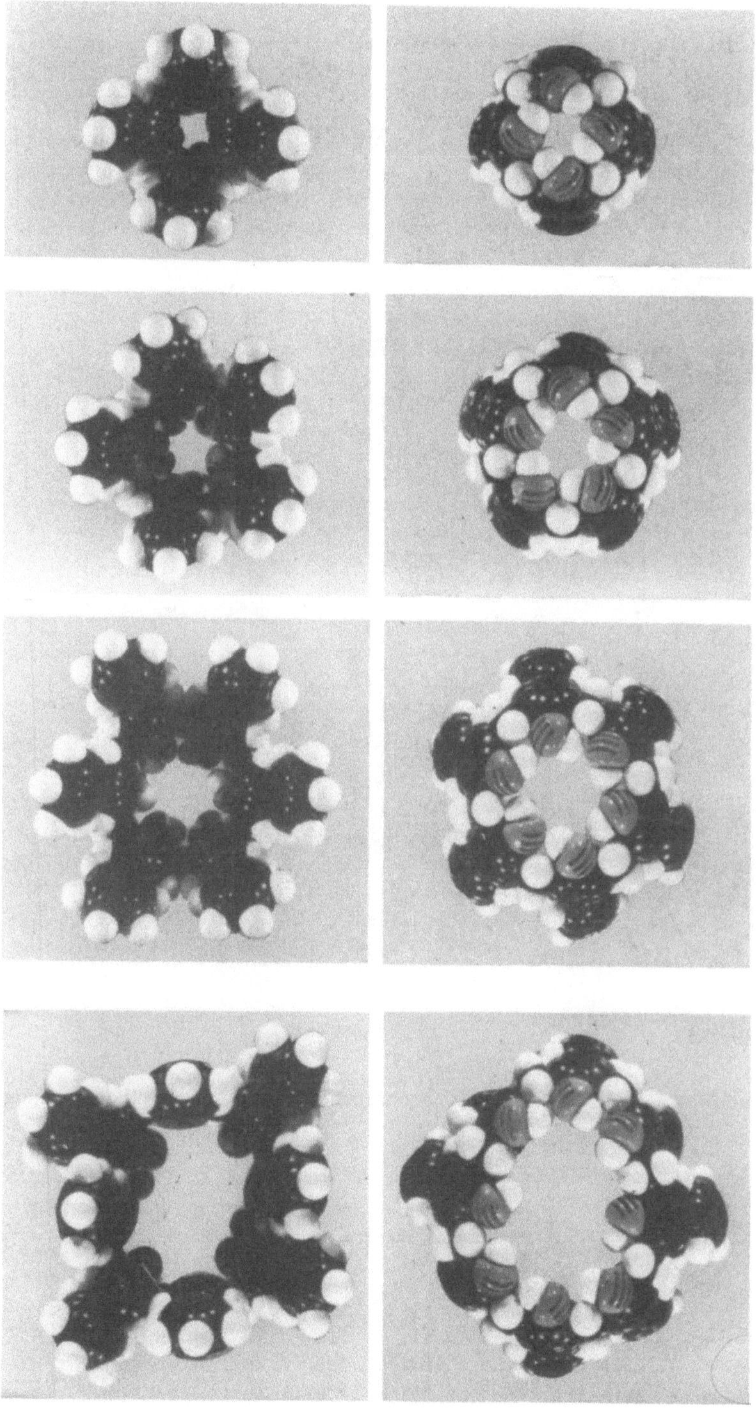

Fig. 12. Space filling molecular models of *p*-phenylcalixarenes. The *p*-phenyl rather than the *p-tert*-butyl substituent has been chosen to more clearly show the proportions of the cavities

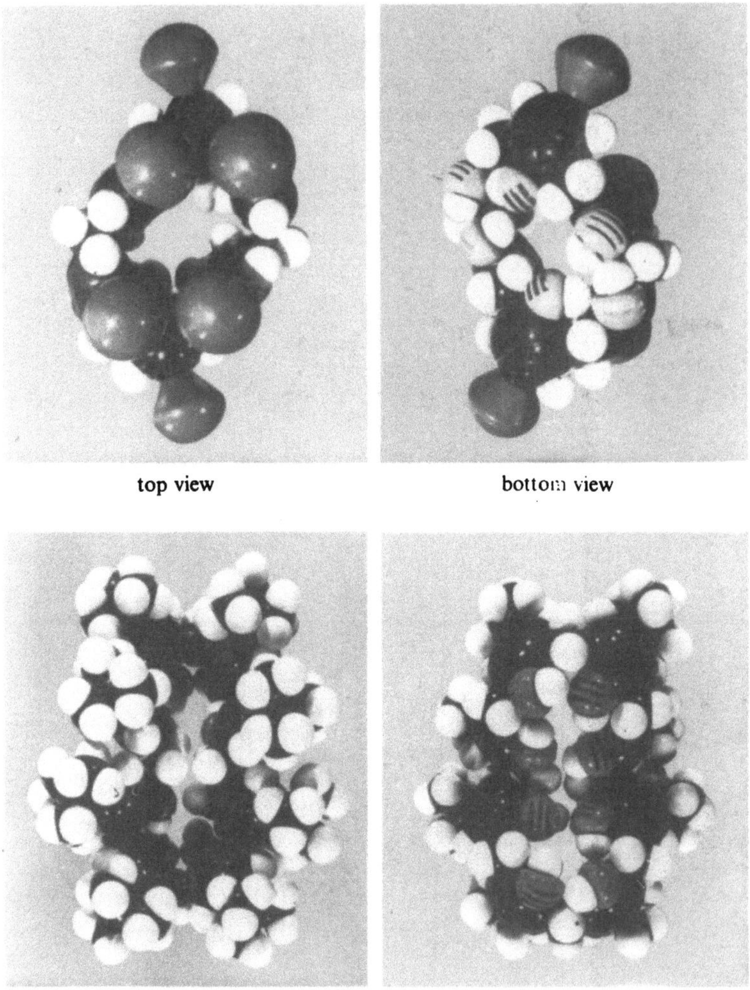

Fig. 13. ϱ-R-Calix[6]arene (top row) and ϱ-R-calix[8]arene (bottom row) in "pinched" conformations

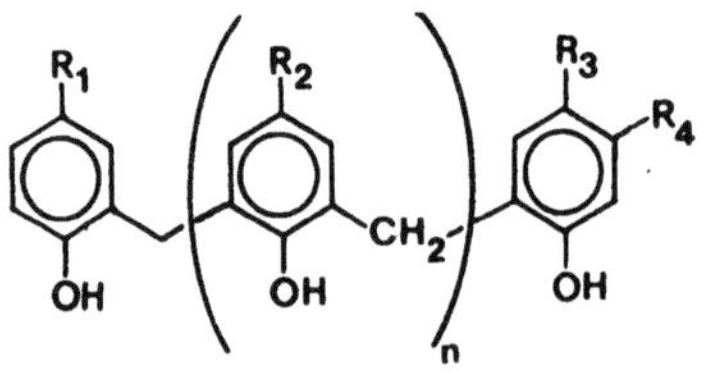

R_1	R_2	R_3	R_4	n	m.p.	Ref.
Me	Me	Me	Me	2	192°	[89]
t-Bu	*t*-Bu	*t*-Bu	Me	2	173°	[89]
Me	Me	Me	Me	3	126°	[93]
Me	Me	Me	H	4	141°	[92]
Me	Me	*t*-Bu	H	4	201°	[92]
Me	Me	*t*-Bu	H	5	210°	[91]

Fig. 14. Melting points of linear oligomers of phenols and formaldehyde

	p-tert-Butyl-calix[4]arene		*p-tert*-Butyl-calix[6]arene		*p-tert*-Butyl-calix[8]arene		Calix[4]arene		Tetraacetyl-*p-tert*-butylcalix[4]arene	
	cold	hot	cold	hot	cold	hot	cold	hot	cold	hot
Cyclohexane	—	—	—	—	—	—	—	±	—	—
Heptane	—	±	—	+	—	—	—	±	—	—
Benzene	±	++	+	+++	+	+	±	++	+	
Toluene	±	++	+	+++	—	+	±	++	+	
Xylenes	—	++	+	+++	—	+	±	++	+	
CH_2Cl_2	±	±	+	++	±	±	+	++	++	
$CHCl_3$	—	±	±	++	+	++	++	+++	++	
CCl_4	—	—	—	—	—	+	±	+	—	—
Ether	—	—	—	—	—	—	—	—	±	±
Tetrahydrofuran	±	—	±	+	±	++	++	+++	++	+++
Diglyme	—	+	+		—	—	±	+	+	++
Dioxane	—	±	—	±	—	+	+	++	+	++
Diphenyl Ether	+	+	±	—	—	—	±	+	+	++
MeOH, EtOH, PrOH and BuOH	—	—	—	—	—	—	—	—	—	+
Acetone	—	—	—	—	—	—	+	++	++	+++
Acetonitrile	—	—	—	—	—	—	—	+	++	+++
Pyridine	—	+	++	++	+	++	+	++	++	
Triethylamine	±	±	++	++	++	++	—	—	—	—
Nitromethane	—	±	—	—	—	±	±	+	++	
Nitrobenzene	±	+	+	+	+	+	+		+	++
DMSO	—	—	—	±	—	±	±	+	±	+
DMF	—	±	—	—	—	—	+	++	+	
Ethyl Acetate	—	+	—	±	—	—	±	+	+	++
Carbon Disulfide	±	+	+++	+++	+++	+++	±	+	—	±
Acetic Acid	—	±	—	±	—	—	—	+	++	
Trifluoroacetic Acid	—	—	—	—	—	—	—	—		
Sulfuric Acid	—	—	—	—	—	—	—	—	+	

Fig. 15. Solubility characteristics of calixarenes. The symbols indicate the following: insoluble (—), very sparingly soluble (±), slightly soluble (+), moderately soluble (++), very soluble (+++)

ing points above 250 °C, in many instances very much higher. For example, *p-tert*-butylcalix[4]arene melts at 344–346 °C, *p-tert*-butylcalix[6]arene melts at 380–381 °C, and *p-tert*-butylcalix[8]arene melts at 411–412 °C; *p*-phenylcalix[4]arene melts at 407–409 °C, and *p*-phenylcalix[8]arene melts above 450 °C. Conversion to calixarene derivatives such as esters and ethers frequently lowers the melting points. For example, the tetramethyl ether of *p-tert*-butylcalix[4]arene melts at 226–228 °C, and the tetrabenzyl ether of *p-tert*-butylcalix[4]arene melts at 230–231 °C. However, the tetraacetate of *p-tert*-butylcalix[4]arene melts at 383–386 °C, and the tetra-trimethylsilyl ether of *p-tert*-butylcalix[6]arene melts at 410–412 °C. In contrast to these high melting points, the acyclic oligomers melt considerably lower, as illustrated by the examples shown in Fig. 14.

4.3 Solubilities of the Calixarenes

A second characteristic feature of the calixarenes is their low solubility in organic solvents. This prevented Zinke from obtaining cryoscopic molecular weights for his cyclic oligomers, and it frequently poses a problem in purifying and characterizing these materials. However, many of the calixarenes have sufficient solubility in $CHCl_3$ to allow osmometric molecular weights to be obtained (concentrations as low as 0.2 g/L can be used [110a]) and in $CDCl_3$ and pyridine-d_5 to make FT-NMR measurements possible. The solubilities of the even-numbered *p-tert*-butylcalixarenes, the unsubstituted calix[4]arene, and the tetraacetate of *p-tert*-butylcalix[4]arene in a wide variety of solvents are recorded in Fig. 15. As would be anticipated, the nature of the *p*-substituent can have a considerable effect on the solubility characteristics of the calixarene. Among the calixarenes substituted with nonpolar groups, the *p*-allylcalixarenes are the most soluble, and the *p*-phenyl and *p*-adamantylcalixarenes are the least soluble. A calix[4]arene substituted with *p*-2-hydroxyethyl groups shows considerable solubility in more polar solvents such as DMSO. Conversion of the calixarenes to the ethers and esters generally increases the solubility in nonpolar solvents. For example, the octamethyl ether of *p*-phenylcalix[8]arene is moderately soluble in $CHCl_3$, in striking contrast to the calixarene itself.

4.4 Infrared Spectra of the Calixarenes

The calixarenes show concentration-independent OH stretching bands in the 3200 cm^{-1} region of the infrared, indicative of very strong intramolecular hydrogen bonding, as illustrated in Fig. 16. This is not uniquely associated with the covalently cyclic nature of the calixarenes, however, for the linear oligomers also show OH stretching bands in the same region, as noted previously (see Sect. 3.1 and Fig. 6). The "fingerprint" regions of the IR spectra of the calixarenes have very similar appearances, but close inspection reveals differences that may be useful in establishing the size of the macrocyclic ring [111]. Thus, the cyclic tetramer is characterized by a moderately strong absorption at 830 cm^{-1}, the cyclic hexamer by absorptions at 750 and 800 cm^{-1}, and the cyclic octamer by the absence of any of these three absorptions. The alkyl ethers of these calixarenes show a unique absorption at 850 cm^{-1} for the cyclic tetramers, a unique absorption at 810 cm^{-1} for the cyclic hexamers, and the absence of these absorptions for the cyclic octamers.

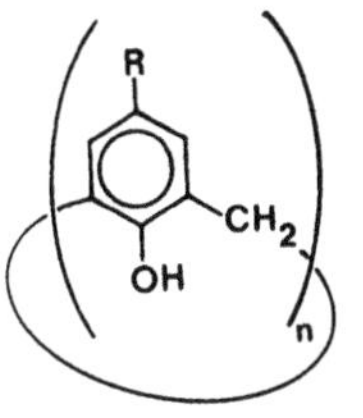

n	$\nu_{OH,\ cm^{-1}}$	1290	1180–1250	990	910	870	830	800	780	750	730	700
		$\nu_{fingerprint,\ cm^{-1}}$										
4	3130–3180	—	3 bands	—	w	w	m	—	m	—	w	w
5	3280											
6	3150–3160	m	3 bands	w	m	m	—	m	—	s	sh	—
7	3155											
8	3230	m	2 bands	w	w	w	—	—	w	—	—	w

Fig. 16. Infrared spectral characteristics of the calixarenes

4.5 Ultraviolet Spectra of the Calixarenes

Kämmerer and Happel [89] compared the UV spectra of several calix[4]arenes with the corresponding linear oligomers and found little difference; both classes of compounds have λ_{max} at ca 280 nm and 288 nm with approximately equal extinction coefficients. As the size of the macrocyclic ring increases, the wavelengths of the maxima remain invariant, although the molar extinction coefficients rise, as shown by the data in Fig. 17. The molar extinction coefficients per aryl ring remain approximately constant for the absorption at 280 nm and increase with ring size for the absorption at 288 nm (e.g. 2075–2225 for n = 4; 4000 for n = 8). The most significant change in the spectra as the ring size increases is in the ratio of the intensities of the 280 and 288 nm absorption bands. For the cyclic tetramer the intensity of the 280 nm band is greater than that of the 288 nm band (ratio ca 1.3), whereas for the cyclic octamer the reverse is true (ratio ca 0.75). For the cyclic pentamer, hexamer, and heptamer the bands are of approximately equal intensity.

R Groups		λ_{max}, mole^{-1} cm^{-1}			
		280 ± 1 nm	288 ± 1 nm	Solvent	Ref.
all Methyl	4	10,500	8,300	Dioxane	89)
all *tert*-Butyl	4	9,800	7,700	$CHCl_3$	110b)
Me and *tert*-Butyl	5	14,030	14,380	Dioxane	93)
all *tert*-Butyl	6	15,500	17,040	$CHCl_3$	110b)
Me and *tert*-Butyl	6	17,210	17,600	Dioxane	92)
all *tert*-Butyl	7	18,200	20,900	$CHCl_3$	110b)
Me and *tert*-Butyl	7	19,800	20,900	Dioxane	91)
all *tert*-Butyl	8	23,100	32,000	$CHCl_3$	110b)

Fig. 17. Ultraviolet absorption characteristics of the calixarenes

4.6 NMR Spectra of the Calixarenes

Among the spectral techniques, that of NMR provides the best indication for the macrocyclic structure of the calixarenes. For example, the ^{1}H NMR spectra of the

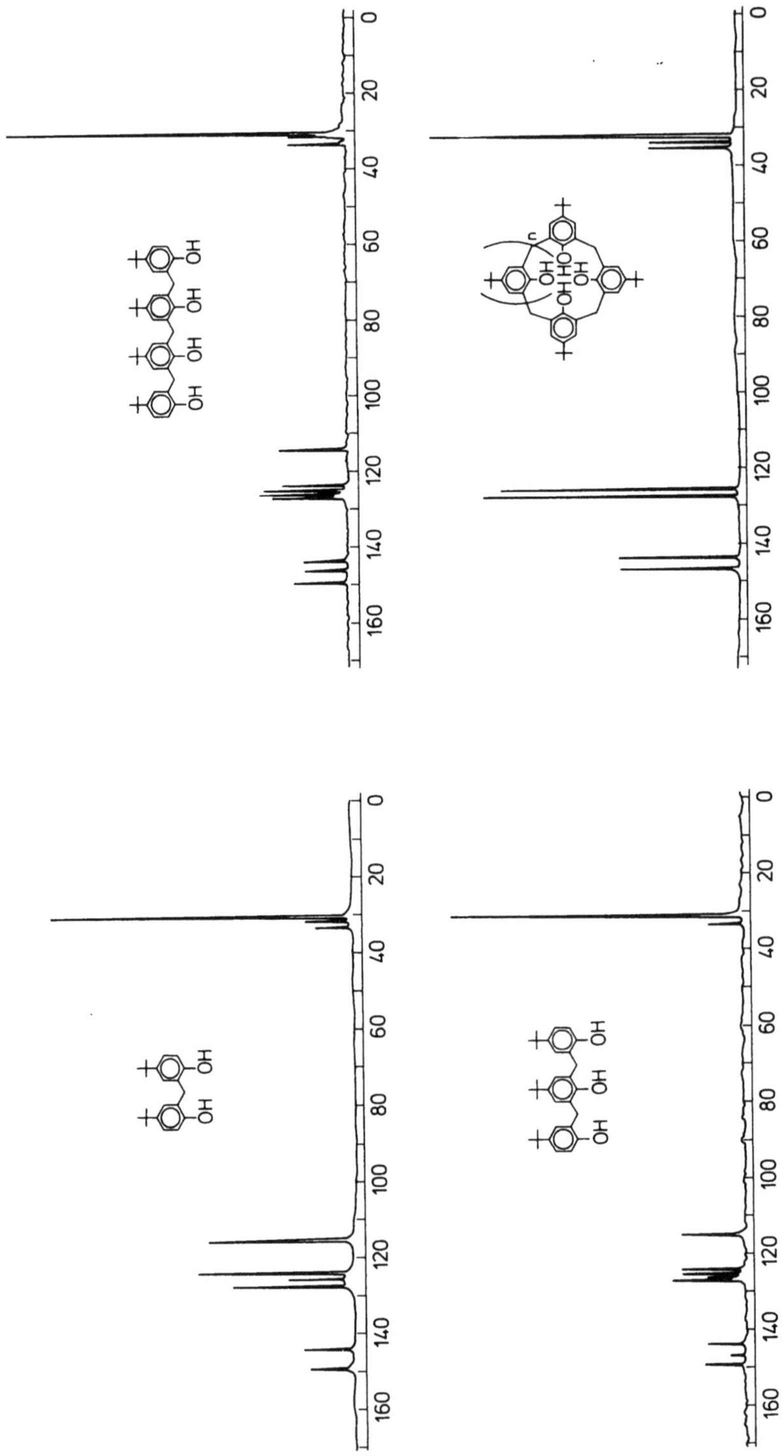

Fig. 18. ^{13}C NMR spectra of linear and cyclic oligomers obtained from the condensation of *p-tert*-butylphenol and formaldehyde

calixarenes in the *p-tert*-butyl series show a single resonance for the aromatic protons, a single resonance for the *tert*-butyl protons, and a temperature-dependent pattern for the methylene protons (discussed in more detail in Sect. 5.1). Shown in Fig. 18 are the ^{13}C NMR spectra of three linear oligomers in the *p-tert*-butyl series along with a cyclic oligomer, illustrating the increasing spectral complexity as the length of the linear oligomer increases but the dramatic decrease in complexity upon cyclization. Due to the symmetry of the cyclic oligomers only four aromatic resonances, one methylene resonance, and two *tert*-butyl resonances are observed.

4.7 Mass Spectra of the Calixarenes

Mass spectra provide a means for determining the molecular weights of the calixarenes via the parent ion signal, but such information must be used with caution. For example, the mass spectrum of *p-tert*-butylcalix[8]arene shows a moderately strong signal at m/e 648, seeming to correspond to a cyclic tetramer. For several years we accepted this as evidence that the major product prepared by the Petrolite procedure was, indeed, the cyclic tetramer. The persistent appearance of small signals at m/e values higher than 648 was disconcerting, however, and when a fully trimethylsilylated sample of the supposed cyclic tetramer was subjected to mass spectral analysis it showed a strong signal at m/e 1872, corresponding to the cyclic octamer. This material showed another quite strong signal at m/e 936, corresponding either to a tetramer resulting from the cracking of the octamer or to the dication of the octamer, thereby providing a possible explanation for the m/e 648 signal from the calixarene itself. In similar vein, mass spectral molecular weights have been reported for calixarenes obtained from *p*-cresol and *p*-methoxyphenol [43] that seem to indicate a cyclic tetramer but that probably arise from compounds of considerably higher molecular weight.

Kämmerer et al. [20,21,89,91–93] have obtained mass spectral data for almost all of the *p*-alkylcalixarenes that they have prepared, including the calix[7]arene *24* which has a parent ion at m/e 883. Kämmerer has stated [94] that the cyclic oligomers show distinct and characteristic differences in the mass spectrum as compared with the analogous linear oligomers, viz. the cyclic oligomers preferentially lose methyl or *tert*-butyl groups and conserve their ring structure, whereas the linear oligomers preferentially cleave into their phenolic units. It has been noted, however, that the calix[5]arene *22a* shows its strongest signal at m/e 480, corresponding to the apparent extrusion of one of the aryl moieties [21], and the cyclic tetramers themselves also show signals corresponding to the extrusion of one, two, and three aryl moieties.

5 Conformational Properties of the Calixarenes

5.1 Conformationally Mobile Calixarenes

The possibility of conformational isomerism in the calix[4]arenes, adumbrated by Megson [112] and Ott and Zinke [113], was made explicit by Cornforth et al. [10] who

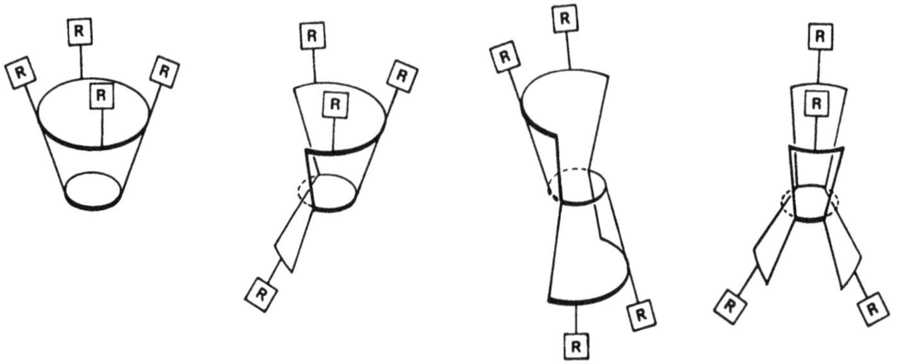

Fig. 19. Conformations of the calix[4]arenes

pointed out that four discrete forms can exist which we refer to [107] as "cone", "partial cone", "1,2-alternate", and "1,3-alternate" conformations, illustrated in Fig. 19. That these are readily interconvertible was first shown by Kämmerer et al. [20, 21] who carried out dynamic ^{1}H NMR studies of *p*-alkylcalix[4]arenes prepared by the stepwise procedure (see Scheme 1). Observing the resonances arising from the $ArCH_2Ar$ methylene hydrogens of the calix, they noted that above room temperature the pattern is a sharp singlet while below room temperature it is a pair of doublets. This can be interpreted in terms of a "cone" conformation that is interconverting rapidly on the NMR time scale at the higher temperature and slowly at the lower temperature. From the coalescence temperature of ca 45 °C it can be calculated that the rate of interconversion is ca 100 sec^{-1}. A similar study was reported in 1977 by Munch [22] using a calixarene prepared by the Petrolite method [47] from *p*-(1,1,3,3-tetramethylbutyl)phenol. Virtually identical results were obtained, and this appeared to provide definite proof (in conjunction with the mass spectral data cited in Sect. 4.7) for the cyclic tetrameric structure of the Munch compound. It seemed inconceivable that the quite rigid calix[4]arene and the much larger, and presumably much more flexible, calix[8]arene might manifest the same dynamic ^{1}H NMR behavior. Yet, this is precisely what occurs. In 1981 Gutsche and Bauer [114] showed that authentic samples of *p-tert*-butylcalix[4]arene and *p-tert*-butylcalix[8]arene do, indeed, show virtually identical dynamic ^{1}H NMR behavior in $CDCl_3$ and C_6D_5Br, as shown in Fig. 20. In pyridine-d_5, however, a difference is observed; whereas the calix[4]arene resolves into a pair of doublets at ca 15 °C, the calix[8]arene shows only a singlet at temperatures as low as —90 °C. The similarity of the dynamic ^{1}H NMR spectra in $CDCl_3$ or C_6D_5Br (nonpolar solvents) is ascribed to intramolecular hydrogen bonding in the cyclic octamer which creates a "pinched" conformation (see Fig. 13) that has the superficial aspect of a pair of tetramers "stuck" together, making the cyclic octamer behave, quite fortuitously, as though it were a cyclic tetramer. The type of hydrogen bonding present in the cyclodextrins has been referred to as "circular" by Saenger [115] who has suggested that circularity confers a special stability on such systems. The cyclic octamer, by "pinching"

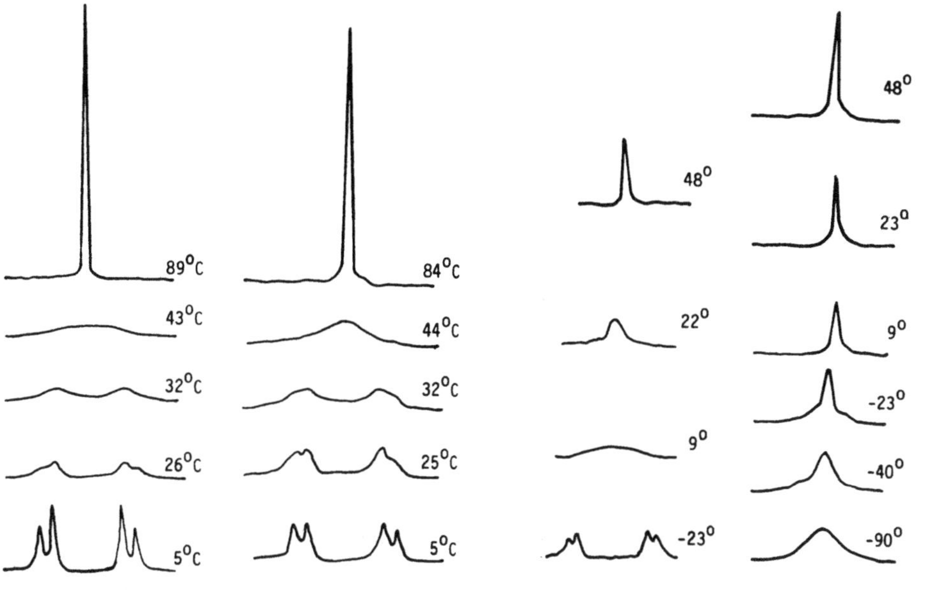

Fig. 20. ^1H NMR spectra of *p-tert*-butylcalix[4]arene and *p-tert*-butylcalix[8]arene at various temperatures in $CDCl_3$ and pyridine-d_5

transannularly, can assume a conformation containing two such circularly hydrogen bonded arrays, each containing four OH groups [114]. A necessary corollary of this postulate, however, is that a pseudorotation is operative which averages the CH_2 *groups* at a rate that is fast on the NMR time scale at the temperatures at which the dynamic NMR studies are conducted, *i.e.*

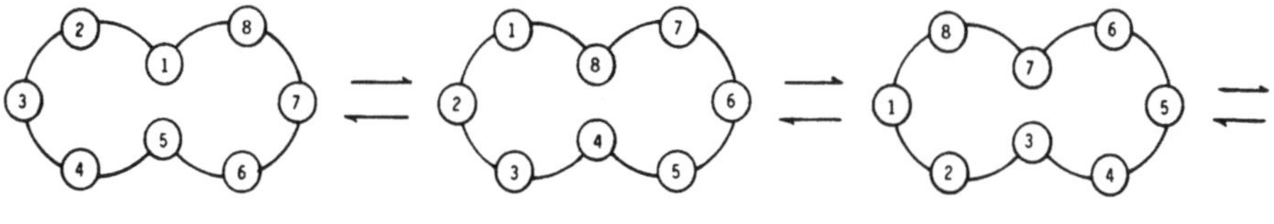

In a fashion similar to the calix[8]arenes, the calix[6]arenes are postulated on the basis of their dynamic ^1H NMR behavior to exist in "pinched" conformations (see Fig. 13) in nonpolar solvents. Transannular "pinching" establishes two circular arrays each containing three intramolecularly hydrogen bonded OH groups. The result is to force the system into a conformation in which the aryl groups in the 1- and 4-positions project outward like a pair of wings perpendicular to the aryl groups in the 2-, 3-, 5-, and 6-positions which are colinear, forming a channel into which other molecules might comfortably fit.

Inversion Rate a Function of:	*p*-Substituents	Ring Size	Solvent	ΔG*, kcal/mole	T°C	Ref.
Ring Size	four t-Bu	4	$CDCl_3$	15.3	52	114)
	five Me	5	$CDCl_3$	12.1		93)
	six t-Bu	6•	$CDCl_3$	13.0	8	116)
	one t-Bu; six Me	7	$CDCl_3$	12.25		91)
	eight t-Bu	8	$CDCl_3$	15.3	53	114)
Solvent	four t-Bu	4	C_6D_5Br	15.2	44	114)
	four t-Bu	4	C_5D_5N	13.4	15	114)
	four H	4	$CDCl_3$	14.6	36	116)
	four H	4	$(CD_3)_2CO$	13.3	−5	116)
	four H	4	CD_3CN	13.3	0	116)
	four H	4	$(CD_3)_2SO$		20	116)
	four C_6H_5	4	$CDCl_3$	15.4	44	116)
	four C_6H_5	4	C_6D_5Br	14.8	36	116)
	four C_6H_5	4	$(CD_3)_2CO$	13.9	8	116)
	four C_6H_5	4	C_5D_5N	12.4	−2	116)
	eight t-Bu	8	C_6D_5Br	15.2	43	114)
	eight t-Bu	8	C_5D_5N	9	< -90	114)
Substituents	four H	4	$CDCl_3$	14.6	36	116)
	three Me; one t-Bu	4	$CDCl_3$	15.9		21)
	one Me; three t-Bu	4	$CDCl_3$	15.4		93)
	four i-Pr	4	$CDCl_3$	14.8	33	116)
	four t-Bu	4	$CDCl_3$	15.2	44	114)
	four t-C_5H_{11}	4	$CDCl_3$	14.5	27	116)
	four $CH_2CH{=}CH_2$	4	$CDCl_3$	15.0	37	116)
	four t-C_8H_{17}	4	$CDCl_3$	14.7	30	116)
	four C_6H_5	4	$CDCl_3$	15.4	44	116)
	two t-Bu; two Cl	4	$CDCl_3$	15.0	38	116)
	eight t-Bu	8	$CDCl_3$	15.3	53	114)
	eight t-C_8H_{17}	8	$CDCl_3$	15.1	53	116)

Fig. 21. Free energies of activation for the conformational inversion of calixarenes

A number of calixarenes with various *p*-substituents and ring sizes have now been studied by means of dynamic ^1H NMR, and the free energies of activation for the conformational inversion have been calculated, as shown in Fig. 21.

5.2 Conformationally Immobile Calixarenes

Cram has defined a "cavitand" as a *synthetic* compound containing an "enforced cavity" [117] large enough to engulf ions or other molecules [118]. The calix[4]arenes are synthetic compounds, and they have the capacity to entrap other molecules. They do not have *enforced* cavities, however, because they are conformationally mobile. To make them cavitands it is necessary to "fix" them either in a "cone" or "partial cone" conformation. One way for accomplishing this is to replace the hydrogens of the OH functions with larger groups. When transformation between conformations occurs an aryl group rotates around the C-2/C-6 axis in a direction that brings the OH groups through the center of the macrocyclic ring. Space filling models indicate that while the OH groups provide little hindrance to this process, larger groups can make the transformation virtually impossible. Ungaro and coworkers [106] and Gutsche and coworkers [107] have studied the process of conformational fixing by this means in some detail.

Each of the conformations for the calix[4]arenes (see Fig. 19) has a distinctive pattern of resonances in the ^{1}H NMR [107], allowing easy structure assignment of the conformationally "fixed" derivatives of the calix[4]arenes. Acetylation of *47* yields a tetraacetate (*50f*) which, by means of ^{1}H NMR, can be shown to be fixed in the "1,3-alternate" conformation [107]. This result was not surprising, because it had already been established that the octahydroxycalix[4]arenes of structure *13* are fixed in a "1,3-alternate-like" conformation [69,70,72]. It appears, however, that the "1,3-alternate" conformation is rather rare among the derivatives of the tetrahydroxycalix[4]arenes (*e.g.* *47–49*) and that the "partial cone" and "cone" conformations are generally favored. For example, *48* yields a tetraacetate (*51f*) that is fixed in a "partial cone" conformation, a result that has been verified by x-ray crystallography [119]. Methylation and ethylation of *47* and *48* and allylation of *47* yield the corresponding ethers *50a*, *50b*, *50c*, *51a* and *51b* in the "partial cone" conformation [107]. Somewhat surprisingly, the tetramethyl ethers *50a* and *51a* are more conformationally mobile than is predicted by inspection of CPK models. At room temperature the ^{1}H NMR spectra of these compounds show only a broad resonance for the methylene hydrogens, quite analogous to the parent calixarenes. These sharpen to simple patterns at elevated temperatures and to more complex patterns at lower temperatures, commensurate with a "partial cone" conformation, and this provides a good example of the fact that solid, space-filling models tend to overestimate the barriers to conformational changes.

"Cone" conformations are established when *48* is converted to its tetra-allyl ether *51c*, benzyl ether *51d*, or trimethylsilyl ether *51e*, when *47* is converted to its

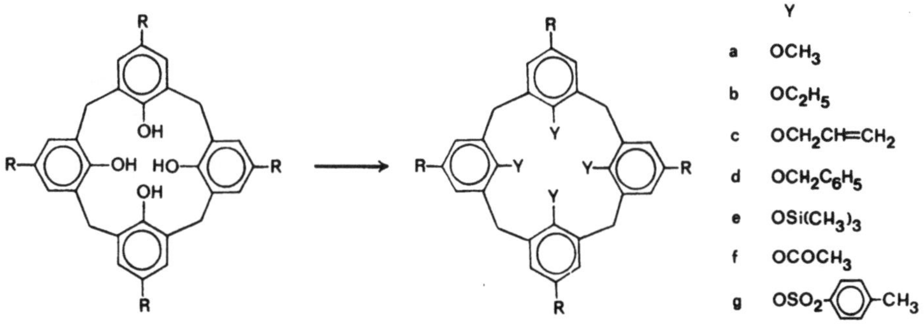

47 (R = H)
48 (R = t-C_4H_9)
49 (R = $CH_2CH{=}CH_2$)

50 (R = H)
51 (R = t-C_4H_9)
52 (R = $CH_2CH{=}CH_2$)

Fig. 22. Calixarenes, calixarene esters, and calixarene ethers tetrabenzyl ether *50d* or tetra-tosylate *50g*, and when *49* is converted to its trimethylsilyl ether *52e* or tetra-tosylate *52g*. In only a few instances have mixtures of conformational isomers been obtained upon derivatization as, for example, in the acetylation of *41* (R = *tert*-Bu) which yields a "1,3-alternate" and two "partial cone" conformers [107, 120]. The "cone" conformation of *41* is established as the only one produced when conversion to the trimethylsilyl derivative occurs.

By appropriate choice of derivatizing agent calix[4]arenes can be fixed either in the "cone" or "partial cone" conformation. Thus, acetylation appears to favor the latter in most cases, and benzylation and trimethylsilylation favor the former. This represents a particularly useful facet of calixarene behavior, for not only does it bring the calixarenes into the family of cavitands but it provides a means for contouring the cavity by design. The explanation for the different conformational outcomes upon derivatization is not known, and the roles played by the derivatizing agent and the reaction conditions remain to be determined. It is clear, however, that the derivatizing agent not only can play a part in determining the conformation of the product but its structure as well, as discussed in Section 3.5. For the partially alkylated calixarenes a "flattened partial cone" conformation is assigned to the trimethyl ether and a "flattened 1,3-alternate" conformation to the monomethyl and dibenzyl ethers. The trimethyl and monomethyl ethers are conformationally less mobile than the tetramethyl ethers, a result that is ascribed to intramolecular hydrogen bonding arising from the free hydroxyl group(s) in the molecules.

Although the calix[8]arenes resemble the calix[4]arenes with respect to conformational mobility, their ester and ether derivatives show no conformational fixation, even at very low temperatures. This difference in behavior arises from the differences in the sizes of the annuli in the two series, the calix[4]arenes having a very small annulus through which groups only as small as OH and OMe can pass but the calix[8]arenes having a much larger one through which groups even as large as tosyl can pass. The calix[6]arenes occupy an intermediate position. Their derivatives are more conformationally mobile than those of the calix[4]arenes but can be frozen out (on the NMR time scale) at low temperatures. Dradi, Pochini and Ungaro [121]

13 →

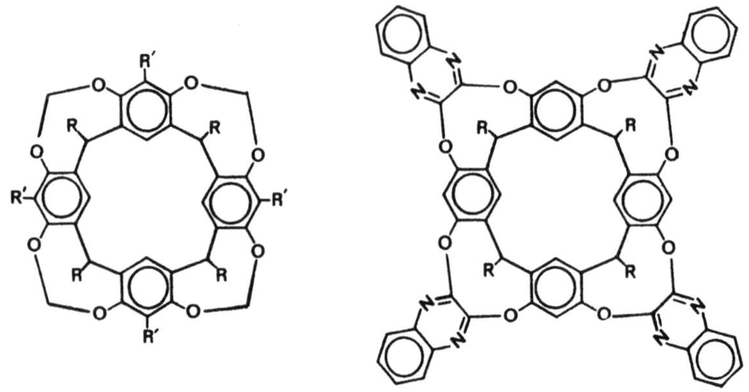

53 *54*

Fig. 23. Synthesis of a rigid, fully bridged calixarene

have studied the dynamic ^{1}H NMR behavior of the hexa-(2-methoxyethyl)ether of *p-tert*-butylcalix[6]arene and conclude that the preferred conformation(s) is either a "1,3,5-alternate" (*i.e.* three "up" and three "down") and/or a "1,4-alternate" (*i.e.* a "winged" conformation similar to that suggested for the free calix[6]arene (see Sect. 5.1). An x-ray crystallographic determination of the compound [33] pictures it in the solid state as an "up, down, out, down, up, out" conformation.

Another way to reduce the conformational mobility of the calixarenes is to bridge two or more regions of the molecule with an appropriate number of atoms. This approach has been successfully applied by Cram [118] to the octahydroxycalix[4]-arene *13* which has been converted to the conformationally rigid compounds *53* (R=H) and *54* by treatment with $ClCH_2Br$ and 2,3-dichloro-1,4-diazanaphthalene, respectively. Working in the tetrahydroxycalix[4]arene series, Gutsche and Bailey [122] have synthesized the dimer unit *55* and have incorporated it, using the stepwise synthesis depicted in Scheme 5, into a dihomooxacalix[4]arene (*57*). Although the single bridges in *56* and *57* do not completely curtail conformational flexing, as do the four bridges in *53* and *54*, complete inversion is no longer possible, and this is manifested in higher coalescence temperatures in the dynamic ^{1}H NMR spectra.

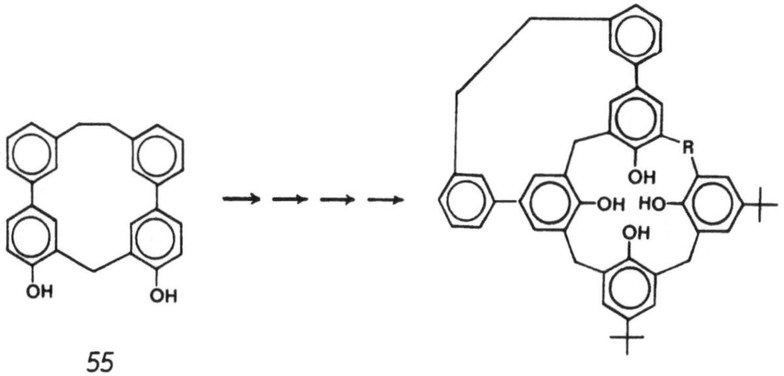

55

56 (R = CH_2)

57 (R = CH_2OCH_2)

Fig. 24. Synthesis of a semi-flexible calix[4]arene

6 Functionalized Calixarenes

One of the principal reasons for the interest in the calixarenes is their potential for serving as enzyme mimics. If they are to operate in this capacity, however, it is necessary that they carry functional groups of various types. The *p-tert*-butylcalix-arenes have proved to be excellent precursors for functional group introduction, for the *p-tert*-butyl group is easily removed by *trans*-alkylation [123]. As outlined in Scheme 7, de-*tert*-butylation of *p-tert*-butylcalix[4]arene (*58*) yields compound *59* which is amenable to functionalization in the *p*-position. For example, bromination of the tetramethyl ether of calix[4]arene (*60*) proceeds smoothly, affording the tetra-bromo compound *61* [124] which, by lithiation followed by carbonation yields the tetramethyl ether of *p*-carboxycalix[4]arene (*65a*) or by cyanation yields the tetramethyl ether of *p*-cyanocalix[4]arene (*65b*) [46]. In similar fashion, Friedel-Crafts acylation of *60* yields *p*-acetylcalix[4]arene, amenable to transformation to *65a* by a haloform oxidation [125].

The bromination and acetylation reactions show that the calixarenes are capable of undergoing conventional electrophilic substitution reactions without rupture of the macrocyclic ring. Zinke et al. [18] reported the nitration of *p-tert*-butylcalixarene (ring size uncertain) and obtained a material that failed to melt up to 400 °C, exploded when heated more strongly, and had a nitrogen analysis in agreement with a tetranitro-calixarene. Attempts to obtain a nitration product from calix[4]arene *59*,

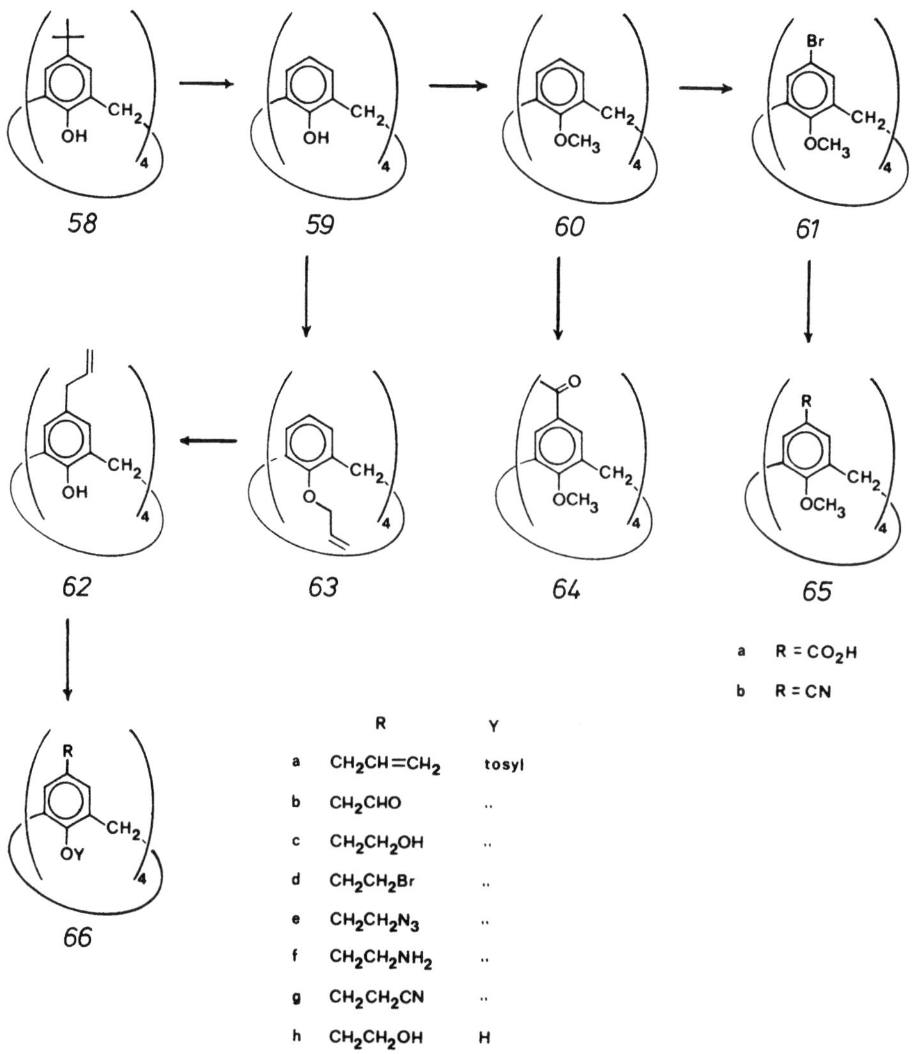

Scheme 7. Synthesis of functionalized calix[4]arenes.

however, have not been successful, a result that is in agreement with Högberg's finding that the octahydroxycalix[4]arenes *13* undergo nitrodealkylation to yield lower molecular weight nitration products [73,12].

As an alternative to electrophilic substitution as a means for introducing functional groups into the calixarenes, a reaction sequence has been developed that involves the conversion of calix[4]arene (*59*) to the tetraallyl ether *63*. When *63* is heated in diethylaniline it undergoes a four-fold *p*-Claisen rearrangement to afford *p*-allylcalix[4]arene (*62*) in excellent yield [126]. From the tetra-tosyl ester of *62* (i.e. compound *66a*) a variety of functionalized calixarenes have been obtained, including the aldehyde *66b*, alcohol *66c*, bromide *66d*, azide *66e*, amine *66f*, and nitrile *66g*. Removal of the tosyl group occurs under mildly basic conditions to yield, for example, *p*-(2-hydroxyethyl)calix[4]arene (*66h*).

Reaction sequences similar to those portrayed in Scheme 7 should also be applicable to the other calixarenes obtained by de-*tert*-butylation of *67* (n = 5, 6, 7, 8), as depicted in Scheme 8. Some of these transformations have already been demonstrated [125], while others remain to be investigated. De-*tert*-butylation of *p-tert*-butylcalix[6]-arene (*67*, n = 6) and *p-tert*-butylcalix[8]arene (*67*, n = 8) proceed fairly smoothly to afford the corresponding calix[6]arene (*68*, n = 6) and calix[8]arene (*68*, n = 8). Conversion of *68* (n = 6) to the hexa-allyl ether (*69b*, n = 6) followed by Claisen rearrangement produces a modest yield of *p*-allylcalix[6]arene (*70*, n = 6). Similar reactions have been shown to take place in the calix[8]arene series as well [125], although the yields are much lower and the products have not yet been fully characterized.

The cavitand *53* (R′ = Br) provides another precursor for the preparation of a functionalized calixarene. Bromination of the octahydroxycalixarene *13* prior to the

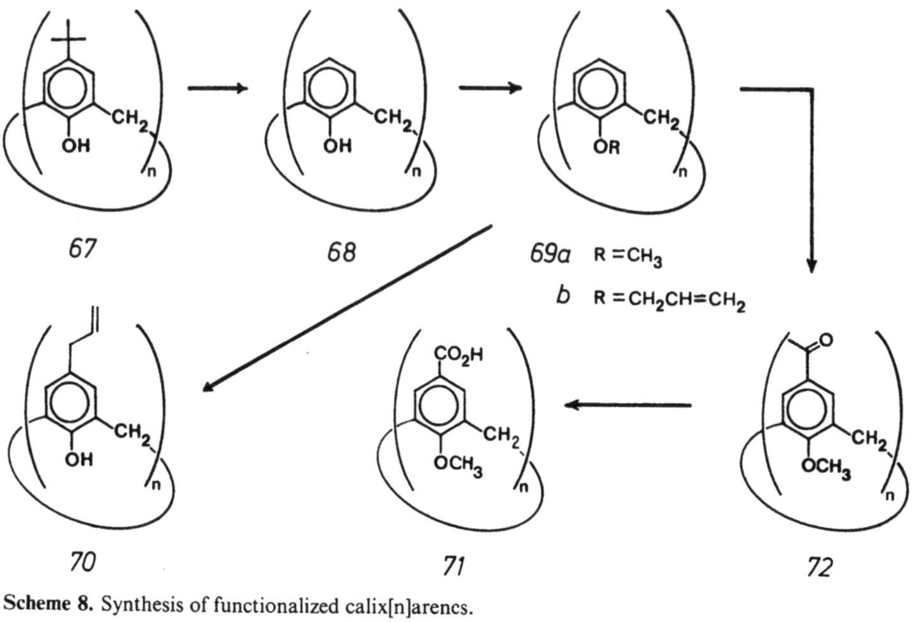

Scheme 8. Synthesis of functionalized calix[n]arencs.

introduction of the methylene bridges between the oxygens affords *53* (R′ = Br), and lithiation followed by treatment with carbon dioxide then yields the tetracarboxy compound *73* [118].

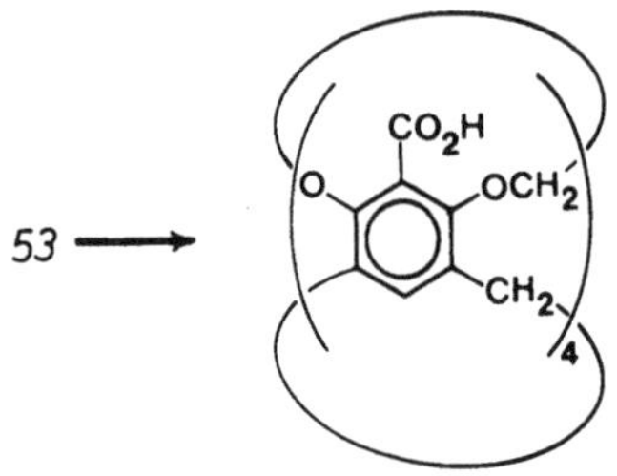

73

7 Complex Formation Involving Calixarenes

7.1 Solid State Complexes with Neutral Molecules

It has long been known that many of the calixarenes retain the solvent from which they are crystallized. For example, *p-tert*-butylcalix[4]arene forms solid state complexes with chloroform, benzene, toluene, xylene, and anisole [32]; *p-tert*-butylcalix[5]arene forms complexes with isopropyl alcohol [26] and acetone [32]; *p-tert*-butylcalix[6]arene forms a complex containing chloroform and methanol [23]; *p-tert*-butylcalix[8]arene forms a complex with chloroform [23]; *p-tert*-butyldihomooxacalix[4]arene forms a complex with methylene chloride [23]. The tenacity with which the guest compound is held varies widely within this group, however. Whereas the cyclic octamer loses chloroform upon standing a few minutes at room temperature and atmospheric pressure, the cyclic hexamer retains some chloroform even after heating for six days at 257 °C at 1 mm pressure. Also, the location of the guest molecule varies, depending on the guest and host, as demonstrated by the x-ray crystallographic work of Andreetti et al. An x-ray crystallographic determination of the *p-tert*-butylcalix[4]-arene-toluene complex shows the calixarene to be in the "cone" conformation and the toluene to be located in the center of the calix [29], *i.e.* an "*endo*-calix" complex as illustrated in Fig. 25. Benzene, *p*-xylene, and anisole are stated to form similar types of complexes [32]. *p*-(1,1,3,3-Tetramethylbutyl)calix[4]arene (*4*, R = 1,1,3,3-tetramethyl-butyl), on the other hand, forms *exo*-calix complexes (channel complexes [127]) with aromatic molecules [128], and calix[5]arene (*5*, R = H) forms a 1:2 complex with acetone in which one acetone molecule is in the calix and the second acetone molecule is external to the calix [32], as illustrated in Fig. 25. An x-ray crystallographic examination of solvent-free *p*-(1,1,3,3-tetramethylbutyl)calix[4]arene [30] shows that the calix-arene is in the "cone" conformation and that two of the 1,1,3,3-tetramethylbutyl groups are oriented inwards to cover the calix, *i.e.* intramolecular complexation. Calix[4]arene, the parent compound lacking *p*-alkyl substituents, also fails to form a tight *endo*-calix complex with aromatic compounds, leading to the postulate [30] that the *tert*-butyl group plays a special role in promoting complexation because

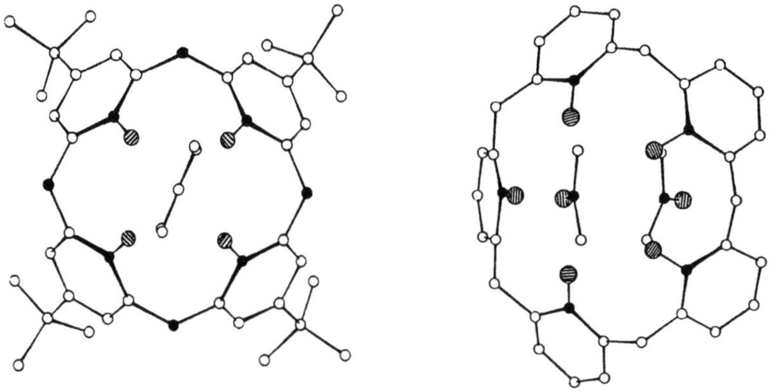

Fig. 25. x-Ray crystallographic structures of 1:1 complex of *p-tert*-butylcalix[4]arene and toluene (left) and 1:2 complex of calix[5]arene and acetone (right)

(a) it is not flexible enough to bend inward and fill the calix itself and (b) it provides CH_3/π interactions [129] between its methyl groups and the aromatic ring of the guest molecule. As has been noted in other studies of clathrate formation [130, 131], subtle changes in the structure of the host and guest can significantly alter the stability and the structure of the host-guest complex.

Selective complexation has been demonstrated [132] by crystallizing *p-tert*-butylcalix[4]arene from 50:50 mixtures of two guest molecules such as benzene and *p*-xylene. It was found that anisole and *p*-xylene are complexed in preference to most other simple aromatic hydrocarbons. *p-tert*-Butylcalix[4]arene has been shown to have some selectivity for mixtures of aromatic hydrocarbons when used as the immobile phase in a chromatographic column [133].

7.2 Solution Complexes with Neutral Molecules

The isolation and characterization of solid state complexes does not necessarily indicate that similar complexes exist in solution. In fact, there are no published data in support of solution complexes of calixarenes. Experiments in our laboratory, using UV/Vis and NMR probes, have failed so far to give any convincing evidence for the formation of solution complexes by various *p*-alkylated-calix[4]arenes, calix[6]arenes, and calix[8]arenes, using chloroform, benzene, toluene, bromobenzene, and a variety of other putative guests. However, certain amines do appear to form such complexes. For example, when *p*-allylcalix[4]arene and *tert*-butylamine are mixed in an acetonitrile solution, the ^1H NMR resonance position of the *tert*-butylamine hydrogens shifts downfield by 0.3 ppm, the resonances of the aromatic hydrogens of the calixarene shift by 0.4 ppm, the rate of conformational inversion of the calixarene is reduced, the relaxation time (T_1) of the *tert*-butylamine hydrogens is shortened, and nuclear overhauser effects on the protons of the allyl group of the calixarene are observed [116]. These phenomena are interpreted as arising from an initial acid-base reaction between the calixarene and the amine to form the calixarene anion (*74*) and the *tert*-butylammonium cation (*75*) followed by association of the ions. The structure of the ion-pair seems best interpreted as an *endo*-calix complex (*76*) as shown in Fig. 26. In support of the idea that a proton transfer constitutes the first event is the fact that weaker amines, such as aniline do not form complexes and the fact that the ^{1}H NMR shifts in the calixarene and amine are comparable to those observed when the calixarene is treated with a strong base and the amine with a strong acid. In support of the idea that the result is an *endo*-calix complex are the ^{1}H NMR data concerning the conformational inversion rate, the relaxation rate, and the nuclear overhauser effects and the fact that some selectivity with respect to the amine component is manifested. For example, neopentyl amine, which has approximately the same basicity as *tert*-butylamine, forms a considerably less stable complex. This is attributed to the "angular" structure of neopentylamine. Assuming

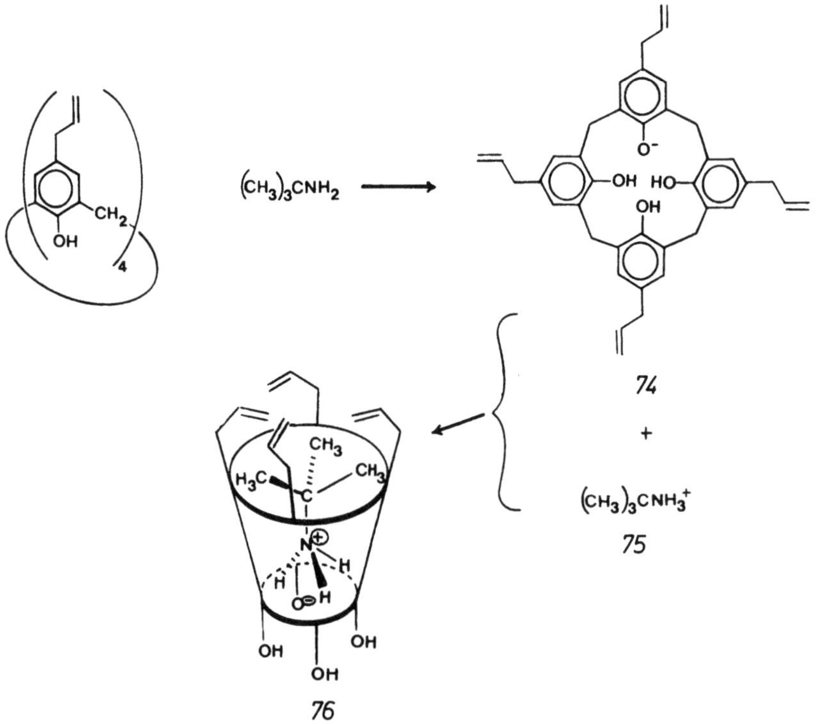

Fig. 26. Complex formation between *p*-allylcalix[4]arene and *tert*-butylamine

that the preferred mode of association between the calixarene and the amine brings the three ammonium hydrogens as close to the calixarene oxygens as possible, the carbon framework of *tert*-butylamine is seen to fit comfortably in the calix, whereas that of neopentylamine presses into the side of the calix. Thus, good evidence is at hand in support of the formation, in solution, of an *endo*-calix complex; but, whether this should be regarded as complexation with a neutral molecule perhaps is debatable.

7.3 Solution Complexes with Cations

The ability of the *p-tert*-butylcalixarenes to transport metal ions through hydrophobic liquid membranes has been studied by Izatt and coworkers [63]. Although the calixarenes are ineffective toward cations in neutral solution, they show considerable transport ability in basic solution, as illustrated by the data in Fig. 27. This contrasts with 18-crown-6 compounds which are far more effective for KNO_3 than KOH. The *p-tert*-butylcalix[4]arene, *p-tert*-calix[6]arene, and *p-tert*-butylcalix[8]-arene all show the same relative cation transport abilities, viz. $Cs^+ > Rb^+ > K^+ > Na^+ \gg Li^+$. The cyclic tetramer shows the greatest selectivity for Cs^+ compared with the other cations, while the cyclic octamer shows the greatest absolute transport ability for Cs^+. Control experiments with *p-tert*-butylphenol, which shows little or no transport ability, indicate that the macrocyclic ring plays a critical role. It seems doubtful that the calixarenes are playing a crown ether-type role by surrounding the cations with a planar array of oxygens. The annulus of the calix[4]arene is too small to do this for an ion as large as Cs^+, and the annulus of the calix[8]arene is too large to do this effectively. The results with the calixarene-amine complexation, discussed in Section 7.2, suggest that the calixarene-metal cation complexes may behave in similar fashion, i.e. that they are *endo*-calix complexes. As pointed

	p-tert-butylcalixarene			18-crown-6	*p-tert*-butylphenol
Source Phase	*4*	*6*	*8*		
LiOH	—	10 ± 1	2.0 ± 0.2	0.9	0.9
NaOH	1.5 ± 0.4	13 ± 2	9 ± 2		
KOH	0.4 ± 0.1	22 ± 3	10.0 ± 0.4		
RbOH	5.6 ± 0.7	71 ± 8	340 ± 20		
CsOH	260 ± 90	810 ± 80	1200 ± 90		

Fig. 27. Cation transport from basic solution by calixarenes, 18-crown-6, and *p-tert*-butylphenol; data given as flux in moles/second-meter^2 × 10^8

out by Izatt et al. [63], cesium ion loses its hydration sphere more easily than the other monovalent cations, and this may be the factor that determines the selectivity favoring cesium. The calix[8]arene in a transannularly "puckered" conformation (see Fig. 13) should have the capacity for carrying two Cs^+ ions per molecule and might be expected to be twice as effective as the calix[4]arene; in fact, it is about four times as effective. The study by Izatt and coworkers indicates that the calixarenes may possess useful features as ion carriers because of their low water solubility, their ability to form *neutral* complexes with cations through loss of a proton, and their potentiality for allowing the coupling of cation transport with the reverse flux of protons.

It has been reported that a complex between the hexa-(methoxyethyl) ether of *p-tert*-butylcalix[6]arene and guanidinium tetraphenyl borate can be detected by ^1H NMR measurements which show large shifts in the resonances of the hydrogens in the methoxyethyl groups and a reduced conformational inversion rate [121]. The complex formed in this instance, however, is different from those formed between the calixarenes and amines or alkaline hydroxides inasmuch as it carries a positive charge and, apparently, involves the methoxyethyl side chains rather than the calix as the site of complexation.

8 Physiological Properties of Calixarenes and Calixarene Derivatives

Phenolic compounds have long been recognized as eliciting various kinds of physiological responses. For example, the urushiols, which are long chain alkyl-substituted catechols, are the active vessicant principle of poison ivy. Somewhat comparable responses, viz. contact dermatitis, have been noted with *p-tert*-butylphenol/formaldehyde resins, particularly the linear tetramer [134]. The *p-tert*-butylcalix[4]arene and *p*-tert-butylcalix[8]arene have been subjected to the Ames' test for mutagenicity [135] and found to give negative readings. Whether this arises from an inherent lack of physiological effect or simply from the great insolubility of these materials is uncertain. Halo-substituted phenols have elicited recurring interest as bacteriostatic agents, and some of this attention has been focused on phenol/formaldehyde oligomers. An early example comes from the Monsanto group [136] who prepared and tested the bacteriostatic properties of linear dimers and trimers from *p*-halophenols and formaldehyde. A recent example is the work of Moshfegh, Hakimelahi et al. [97–99, 137] who have prepared a long series of linear as well as cyclic oligomers from *p*-halophenols and formaldehyde (see Scheme 4) and have tested their *in vitro* activity against various pathogenic organisms. Significant activity was noted for linear tetramers and cyclic tetramers, and it is suggested [97] that this correlates with the chelating ability (*e.g.* toward Fe^{++}) of these compounds.

An extensive medical and biochemical literature has arisen during the last several decades concerning the oxyalkyl derivatives of simple phenols as well as phenol/formaldehyde condensation products. Of particular note in the context of this review is the work of Cornforth and coworkers [10, 108] on the polyoxyethyl derivatives of the *p-tert*-butyl and *p*-(1,1,3,3-tetramethylbutyl)calixarenes. In the earlier work [10] oxyethylation was effected simply by treating the calixarenes with ethylene oxide to give products designated as "macrocyclons". In the later work [108] a more carefully controlled set of experiments was carried out using the high melting compound from the condensation of *p*-(1,1,3,3-tetramethylbutyl)phenol and formaldehyde. Tests of these compounds for tuberculostatic activity led to the conclusion that the lipophilic-hydrophilic balance of the molecule may be the most critical factor but that resistance to chemical breakdown *in vivo* is also important if activity is to be shown against the relatively slow course of experimental tuberculosis. Macrocyclon and other oxyalkylated phenols have been tested against a variety of other organisms [138] and as carcinostatic agents [139].

9 Conclusion

The calixarenes, along with the cyclodextrins, crown ethers, and other macrocyclic compounds, provide an entry into a field of research that has been referred to in an earlier volume of this series as "cavity chemistry"[140]. Much of the work in the field of calixarenes has focused on the synthesis of these compounds. Methods are now available for constructing the basic ring system either by single step processes, which are severely limited with respect to substituent group possibilities, or by multistep processes, which permit greater substituent group flexibility. Methods have been developed for the introduction of functional groups onto the calixarene framework, either at the *p*-positions of the aromatic rings or the oxygens of the phenolic groups. Thus, a variety of cavity-containing calixarenes with various ring sizes, substituents, and conformations are now available for study. The structures of several solid state complexes formed between small molecules and calixarenes have been verified by x-ray crystallography, and the formation of complexes in solution between calixarenes and amines has been demonstrated. It remains yet to demonstrate the catalytic and enzyme mimic properties of the calixarenes, and work is in progress with compounds 66c and 66h (as hydrolysis catalysts)[141], compound 66f (as metal chelator and oxygen carrier)[48], and various other functionalized calixarenes.

10 Acknowledgements

I am indebted to my coworkers, whose names appear in the references of this review article, for their splendid work in calixarene chemistry. Their conscientiousness and ingenuity are primarily responsible for the progress that has been made in our laboratories. Particular thanks are given to Drs. J. H. Munch and F. J. Ludwig of the Petrolite Corporation, whose long and continuing interest and collaboration in this work is greatly appreciated. Finally, I am indebted to Washington University, the National Science Foundation, and the National Institutes of Health for financial support and to the John Simon Guggenheim Memorial Foundation for a fellowship, during the tenure of which the invitation to write thise review was received and a portion of the manuscript completed.

11 References

1. Cram, D. J., Cram, J. M.: Science *183*, 803 (1974)
2. Lehn, J. M.: Pure Applied Chem. *50*, 871 (1978)
3. Gutsche, C. D., Muthukrishnan, R.: J. Org. Chem. *43*, 4905 (1978)
4. Gutsche, C. D.: Accts. Chem. Res. *16*, 161 (1983)
5. Patterson, A. M., Capell, L. T., Walker, D. F.: The Ring Index, 2nd. ed., Amer. Chem. Soc., Washington, D.C., 1960, Ring Index No. 6485
6. Cram, D. J., Steinberg, H.: J. Am. Chem. Soc. *73*, 5691 (1951)
7. IPUAC Tentative Rules for Nomenclature of Organic Chemistry, Section. E. Fundamental Stereochemistry; cf. J. Org. Chem. *35*, 2849 (1970)
8. Zinke, A., Kretz, R., Leggewie, E., Hössinger, K.: Monatsh. Chem. *83*, 1213 (1952)

9. Hayes, B. T., Hunter, R. F.: Chem. Ind., 193 (1956): J. Applied Chem. *8*, 743 (1958)
10. Cornforth, J. W., D'Arcy Hart, P., Nicholls, G. A., Rees, R. J. W., Stock, J. A.: Brit. J. Pharmacol. *10*, 73 (1955)
11. We are indebted to Dr. Kurt L. Loening of Chemical Abstract Services for his guidance in questions concerning the numbering of these compounds, which follows the IUPAC guidelines as noted for compound *2* in Fig. 1
12. Högberg, A. G. S.: PhD Dissertation, Royal Institute of Technology, Stockhom, 1977. We are indebted to Dr. Högberg for making available to us this excellent thesis which suggests the "intraannular" and "extraannular" terminology
13. Beyer, A.: Ber. dtsch. chem. Ges. *5*, 25, 280, 1094 (1872)
14. Numerous accounts of the history of phenol-formaldehyde resins are available, including Baekeland, L. H.: Ind. Eng. Chem. (Industry) *5*, 506 (1913); Carswell, T. S.: Phenoplasts, Interscience Publishers, Inc., New York, 1947; Gillis, J., Oesper, R. E.: J. Chem. Ed. *41*, 224 (1964)
15. Zinke, A., Ziegler, E.: Ber. dtsch. chem. Ges. *74*, 1729 (1941)
16. Zinke, A., Ziegler, E.: ibid. *77B*, 264 (1944)
17. Zinke, A., Zigeuner, G., Hössinger, K., Hoffmann, G.; Monatsh. Chem. *79*, 438 (1948)
18. Zinke, A., Ott, R., Garrana, F. H.: Monatsh. Chem. *89*, 135 (1958)
19. Zinke, A.: J. Appl. Chem. *1*, 257 (1951)
20. Kämmerer, H., Happel, G., Caesar, F.: Makromol. Chem. *162*, 179 (1972)
21. Happel, G., Mathiasch, B., Kämmerer, H.: Makromol. Chem. *176*, 3317 (1975)
22. Munch, J. H.: Makromol. Chem. *178*, 69 (1977)
23. Gutsche, C. D., Dhawan, B., No, K. H., Muthukrishnan, R.: J. Am. Chem. Soc. 103, 3782 (1981)
24. Gutsche, C. D., Muthukrishnan, R., No, K. H.: Tetrahedron Lett. 2213 (1979)
25. Muthukrishnan, R., Gutsche, C. D.: J. Org. Chem. *44*, 3962 (1979)
26. Ninagawa, A., Matsuda, H.: Makromol. Chem. Rapid Comm. *3*, 65 (1982)
27. Nakamoto, Y., Ishida, S.:: ibid. *3*, 705 (1982)
28. Mukoyama, Y., Tanno, T.: Org. Coating & Plastics Chem. *40*, 894 (1979)
29. Andreetti, G. D., Ungaro, R., Pochini, A.: J. Chem. Soc. Chem. Comm. 1005 (1979)
30. Andreetti, G. D., Pochini, A., Ungaro, R.: J. Chem. Soc., Perkin II, 1773 (1983)
31. Ninagawa, A.: private communication
32. Coruzzi, M., Andreetti, G. D., Bocchi, V., Pochini, A., Ungaro, R.: J. Chem. Soc., Perkin II, 1133 (1982)
33. Andreetti, G. D., Ungaro, R., Pochini, A.: Abstr. 2nd Internat. Symp. on Clathrate Compounds and Molecular Inclusion Phenomena, Parma, 1982, p. 88
34. Andreetti, G. D., Ungaro, R., Pochini, A.: J. Chem. Soc. Chem. Comm. 533 (1981)
35. Raschig, F.: Z. für angew. Chem. *25*, 1939 (1912)
36. Baekeland, L. H.: J. Ind. Eng. Chem. (Industry) *5*, 506 (1913)
37. Niederl, J. B., McCoy, J. S.: J. Am. Chem. Soc. *65*, 629 (1943)
38. Koebner, M.: Z. Angew. Chem. *46*, 251 (1933)
39. Finn, S. R., Lewis, G. J.: J. Chem. Soc. Ind. *69*, 132 (1950)
40. Foster, H. M., Hein, D. W.: J. Org. Chem. *26*, 2539 (1961)
41. The *p-tert*-butylcalix[8]arene and *p*-(1,1,3,3-tetramethylbutyl)calix[8]arene have been correlated by removing the alkyl groups from each of these compounds and obtaining the same calix[8]-arene (*8*, R = H)[32]
42. Gutsche, C. D., Kung, T. C., Hsu, M.-L.: Abstr. 11th Midwest Regional Meet. Amer. Chem. Soc., Carbondale, IL, 1975, no. 517
43. Patrick, T. B., Egan, P. A.: J. Org. Chem. *42*, 382 (1977)
44. Patrick and Egan [43] reported molecular weights for their products using either mass spectroscopy or the Rast method (camphor as solvent). However, unless the *absence* of m/e signals higher than that of the anticipated parent ion (i.e. cyclic tetramer) is demonstrated, the mass spectral data can be misleading. The Rast method, requiring the compound to be at least moderately soluble in camphor, would appear to be inapplicable in the case of the cyclic oligomers, all of which are quite insoluble in most organic materials
45. Chen, S. I.; unpublished observations
46. Pagoria, P. F.: unpublished observations
47. Buriks, R. S., Fauke, A. R., Munch, J. H.: U.S. Patent 4,259,464, filed 1976, issued 1981

48. Levine, J. A.: unpublished observations
49. Stewart, D.: unpublished observations
50. Ullman, F., Brittner, K.: Ber. dtsch. chem. Ges. *42*, 2539 (1909)
51. Wohl, A., Mylo, B.: ibid. *45*, 2046 (1912)
52. Hultzsch, K.: ibid. *75B*, 106 (1942)
53. v. Euler, H., Adler, E., Cedwall, J. O., Törngren, O.: Ark. f. Kemi Mineral. Geol. *15A*, No. 11, 1 (1941)
54. Gardner, P. D., Sarrafizadeh, H., Brandon, R. L.: J. Am. Chem. Soc. *81*, 5515 (1959)
55. Wegler, R., Herlinger, H.: Methoden der Organischen Chemie (Houben-Weyl), XIV/2; Makromolecular Stoffe, Thieme Verlag, Stuttgart, 1963, p. 257
56. Evans, D. A., Golob, A. M.: J. Am. Chem. Soc. *97*, 4765 (1975); Thies, R. W., Seitz, E. P.: J. Org. Chem. *43*, 1050 (1978)
57. Adler, E.: Arkiv. f. Kemi Mineral. Geol. *14B*, No. 23, 1 (1941)
58. v. Euler, H., Adler, E., Bergstrom, B.: Arkiv. f. Kemi Geol., *14B*, No. 25, 1 (1941); Hultzsch, K.: Kunststoffe *52*, 19 (1962)
59. Zinke, A., Ziegler, E., Hontschik, I.: Monatsh. Chem. *78*, 317 (1948); Ziegler, E., Hontschik, I.: ibid. *78*, 325 (1948)
60. In addition to water and formaldehyde, other extrusion products, such as benzaldehydes, have also been shown to be formed upon heating mixtures above 150 °C [59)]
61. Cairns, T., Eglinton, G.: Nature *196*, 535 (1962)
62. Dhawan, B., Gutsche, C. D.: J. Org. Chem. *48*, 1536 (1983)
63. Izatt, R. M., Lamb, J. D., Hawkins, R. T., Brown, P. R., Izatt, S. R., Christensen, J. J.: J. Am. Chem. Soc. *105*, 1782 (1983)
64. Burke, W. J., Craven, W. E., Rosenthal, A., Ruetman, S. H., Stephens, C. W., Weatherbee, C.: J. Polymer Sci. *20*, 75 (1956)
65. Dhawan, B.: unpublished observations
66. Michael, A.: Amer. Chem. J. *5*, 338 (1883)
67. Niederl, J. B., Vogel, H. J.: J. Am. Chem. Soc. *62*, 2512 (1940)
68. Erdtman, H., Haglid, F., Ryhage, R.: Acta Chem. Scand. *18*, 1249 (1964)
69. Erdtman, H., Högberg, S., Abrahamsson, S., Nilsson, B.: Tetrahedron Lett. 1679 (1968); Nilsson, B.: Acta. Chem. Scand. *22*, 732 (1968)
70. Palmer, K. J., Wong, R. Y., Jurd, L., Stevens, K.: Acta Crystallogr. *B32*, 847 (1976)
71. Högberg, A. G. S.: J. Org. Chem. *45*, 4498 (1980)
72. Högberg, A. G. S.: J. Am. Chem. Soc. *102*, 6046 (1980)
73. Additional evidence for the acid-lability of the cyclic tetramers (*13*) is seen in their tendency to undergo nitrodealkylation. For example, treatment of *13* (R = Phenyl, R′ = Methyl) with concentrated nitric acid yields 4,6-dinitroresorcinol dimethyl ether and 3,5-dinitrobenzoic acid [12)]
74. Ballard, J. L., Kay, W. B., Kropa, E. L.: J. Paint Technol. *38*, 251 (1966)
75. Bottino, F., Montaudo, G., Maravigna, P.: Ann. Chim. (Rome) *57*, 972 (1967)
76. Brown, W. H., Hutchinson, B. J.: Can. J. Chem. *56*, 617 (1978) and preceding papers back to Brown, W. H., Sawatsky, H.: ibid. *34*, 1147 (1956)
77. Chastrette, M., Chastrette, F.: J. Chem. Soc. Chem. Comm. 534 (1973)
78. Healy, M. de S., Rest, A. J.: J. Chem. Soc. Chem. Comm. 140 (1981)
79. Högberg, A. G. S., Weber, M.: Acta. Chem. Scand. B *37*, 55 (1983)
80. Ahmed, M., Meth-Cohn, O.: Tetrahedron Lett. 1493 (1969); J. Chem. Soc. C 2104 (1971)
81. Rothemund, P., Gage, C. L.: J. Am. Chem. Soc. *77*, 3340 (1955)
82. Collman, J. P., Gagne, R., Reed, C., Halbert, T. R., Lang, G., Robinson, W. T.: J. Am. Chem. Soc. *97*, 1427 (1975)
83. Robinson, G. M.: J. Chem. Soc. *107*, 267 (1915); Collet, A., Gabard, J.: J. Org. Chem. *45*, 5400 (1980) and references therein
84. Freeman, W. A., Mock, W. L., Shih, N.-Y.: J. Am. Chem. Soc. *103*, 7367 (1981)
85. Meth-Cohn, O.: Tetrahedron Lett. 91 (1973)
86. Sawa, N., Nomoto, T., Aida, K., Suzuki, T.: J. Synth. Org. Chem. *33*, 1007 (1975)
87. Bergman, J., Högberg, S., Lindström, J.-O.: Tetrahedron *26*, 3347 (1970); Raverty, W. D., Thomson, R. H., King, T. J.: J. Chem. Soc., Perkin I, 1204 (1977)

88. Hunter, R. F., Turner, C.: Chem. Ind., 72 (1957) also reported the synthesis of a cyclic octamer containing six *m*-bridges and two *p*-bridges, the remaining four *m*-positions being "blocked" with methyl groups
89. Kämmerer, H., Happel, G.: Makromol. Chem. *179*, 1199 (1978)
90. Kämmerer, H., Happel, G., Böhmer, V., Rathay, D.: Monatsh. Chem. *109*, 767 (1978)
91. Kämmerer, H., Happel, G.: Makromol. Chem. *181*, 2049 (1980)
92. Kämmerer, H., Happel, G.: ibid. *112*, 759 (1981)
93. Kämmerer, H., Happel, G., Mathiasch, B.: ibid. *182*, 1685 (1981)
94. Kämmerer, H., Happel, G.: in: Weyerhaeuser Science Symposium on Phenolic Resins, Tacoma, Washington, 1979, p. 143
95. Gutsche, C. D.; No, K. H.: J. Org. Chem. 47, *2708* (1982)
96. Böhmer, V., Chhim, P., Kämmerer, H.: Makromol. Chem. *180*, 2503 (1979)
97. Moshfegh, A. A., Badri, R., Hojjatie, M., Kaviani, M., Naderi, B., Nazmi, A. H., Ramezanian, M., Roospeikar, B., Hakimelahi, G. H.: Helv. Chim. Acta. *65*, 1221 (1982)
98. Moshfegh, A. A., Mazandarani, B., Nahid, A., Hakimelahi, G. H.: Helv. Chim. Acta *65*, 1229 (1982)
99. Moshfegh, A. A., Baladi, E., Radnia, L., Afsanch, S. L. R., Hosseini, A. S., Tofigh, S., Hakimelahi, G. H.: Helv. Chim. Acta 65, 1264 (1982)
100. No, K. H., Gutsche, C. D.: J. Org. Chem. 47, 2713 (1982)
101. Hultzsch, K.: Kunststoffe *52*, 19 (1962)
102. von Euler, H., Adler, E., Bergstrom, B.: Ark. Kemi. Mineral Geol. *14B*, No. 30, 1 (1941)
103. Kämmerer, H., Dahm, M.: KunstPlast (Solothurn Switz.) *6*, 20 (1959)
104. Still, W. C., Kahn, M., Mitra, A.: J. Org. Chem. *43*, 2923 (1978)
105. Kricheldorf, H. R., Kaschig, J.: Liebigs Ann. Chem. 882 (1976)
106. Bocchi, V., Foina, D., Pochini, A., Ungaro, R.: Tetrahedron *38*, 373 (1982)
107. Gutsche, C. D., Dhawan, B., Levine, J. A., No, K. H., Bauer, L. J.: Tetrahedron *39*, 409 (1983)
108. Cornforth, J. W., Morgan, E. D., Potts, K. T., Rees, R. J. W.: Tetrahedron *29*, 1659 (1973)
109. Klebe, J. F., Finkbeiner, H., White, D. M.: J. Am. Chem. Soc. *88*, 3390 (1966)
110. (a) Gutsche, A. E.: unpublished observations; (b) Ludwig, F. J.: unpublished observations
111. We are indebted to Mr. Christopher Cramer for devising a computer program for comparing the "fingerprint" region data for a variety of calixarenes
112. Megson, N. R. L.: Oesterr. Chem. Z. *54*, 317 (1953)
113. Ott, R., Zinke, A.: Oesterr. Chem. Z. *55*, 156 (1954)
114. Gutsche, C. D., Bauer, L. J.: Tetrahedron Lett. 4763 (1981)
115. Saenger, W., Betzel, C., Brown, G. M.: Angew. Chem. Int. Ed. Engl. *22*, 883 (1983); Saenger, W., Betzel, C., Hingerty, B., Brown, G. M.: Nature 296, 581 (1982); Saenger, W.: ibid, 279, 343 (1979)
116. Bauer, L. J.: unpublished observations
117. Helgeson, R. C., Mazaleyrat, J.-P., Cram, D. J.: J. Am. Chem. Soc. *103*, 3929 (1981)
118. Moran, J. R., Karbach, S., Cram, D. J.: J. Am. Chem. Soc. *104*, 5826 (1982); Cram, D. J.: Science *219*, 1177 (1983)
119. Rizzoli, C., Andreetti, G. D., Ungaro, R., Pochini, A.: J. Molec. Structure *82*, 133 (1982)
120. The formation of two "partial cone" compounds is the result of the reduced symmetry of *41* (R = *tert*-Bu) which gives rise to six conformers; viz. one "cone", two "partial cone", two "1,2-alternate", and one "1,3-alternate"
121. Dradi, E., Pochini, A., Ungaro, R.: Abstr. 2nd Internat. Symposium on Clathrate Compounds and Molecular Inclusion Phenomena, Parma/Italy, 1982, p. 84
122. Bailey, D. W.: unpublished observations
123. See Tashiro, M.: Synthesis, 921 (1979) for general references and see Böhmer, V., Rathay, D., Kämmerer, H.: Org. Prep. Proc. Int. *10*, 113 (1978) for a closely analogous example
124. Compound *61*, which melts at 248–250 °C, can be demethylated with BBr_3 to *p*-bromocalix[4]-arene, which melts above 430 °C. The structure of *p*-bromocalix[4]arene has been established by the criteria discussed in this article except x-ray crystallography. *p*-Chlorocalix[4]arene, claimed to be the product of the synthesis illustrated in Scheme 4 [97], is reported to melt at 239–242 °C. The enormous difference in melting points between the *p*-bromo and *p*-chloro compounds casts considerable doubt on the structure of the *p*-chloro compound.
125. Lin, L. G.: unpublished observations

126. Gutsche, C. D., Levine, J. A.: J. Am. Chem. Soc. *104*, 2652 (1982)
127. Hagan, M.: Clathrate Inclusion Compounds, Reinhold, New York, 1962
128. Andreetti, G. D.: Internat. Symp. on Clathrate Compounds and Molecular Inclusion Phenomena, Jackranka/Poland, 1980; quoted in ref. [32]
129. Uzawa, J., Zushi, S., Kodama, Y., Fukuda, Y., Nishihata, K., Umemura, A., Nishio, M., Hirota, M.: Bull. Chem. Soc. Japan *53*, 3623 (1980)
130. MacNicol, D. D., McKendrick, J. J., Wilson, D. R.: Chem. Soc. Rev. *7*, 65 (1978)
131. Hilgenfeld, R., Saenger, W.: Angew. Chem. Suppl. 1690 (1982)
132. Andreetti, G. D., Mangia, A., Pochini, A., Ungaro, R.: Abstr. 2nd Internat. Symposium on Clathrate Compounds and Molecular Inclusion Phenomena, Parma/Italy, 1982, p. 42
133. Smolková-Keulemansová, E., Feltl, L.: ibid., p. 45
134. Schubert, H., Agatha, G.: Dermatosen im Beruf und Umwelt *27*, 49 (1979)
135. We are indebted to Professor Barry Commoner for making available the testing facilities of the Center for the Study of Biological Systems, Washington University, St Louis, Mo.
136. Beaver, D. J., Shumard, R. S., Stoffel, P. J.: J. Am. Chem. Soc. 75, 5579 (1953)
137. Hakimelahi, G. H., Moshfegh, A. A.: Helv. Chim. Acta *64*, 599 (1981)
138. A representative but not inclusive list of references includes: (a) D'Arcy Hart, P., Payne, S. N.: Brit. J. Pharmacol. *43*, 190 (1971); (b) D'Arcy Hart, P., Gordon, A. H.: Nature *222*, 672 (1969); (c) Niffenegger, J., Youmans, G. P.: Brit. J. Exptl. Pathol. *41*, 403 (1960); (d) Fulton, J. D.: Nature 187, 1129 (1960); (e) Depamphilis, M. L.: J. Bact., *105*, 1184 (1971); (f) Allwood, M. C.: Microbios *7*, 209 (1973)
139. A representative but non-inclusive list of references includes: (a) Carter, R. L., Birbeck, M. S. C., Stock, J. A.: Int. J. Cancer *7*, 32 (1971); (b) Miller, G. W., Janicki, B. W.: Cancer Chem. *52*, 243 (1968); (c) Franchi, G., Morasca, L., Reyers-Delgi-Innocenti, I., Garattini, S.: Europ. J. Cancer *7*, 535 (1971); (d) Rosso, R., Donelli, M. G., Franchi, G., Garattini, S.: Cancer Chemotherapy Rpt. *54*, 79 (1970)
140. Vögtle, F., Sieger, H., Müller, W. M.: Host Guest Complex Chemistry I, in Topics in Current Chemistry, Springer-Verlag, Berlin, 1981, p. 107
141. Sujeeth, P. K.: unpublished observations

Complexation of Iron by Siderophores
A Review of Their Solution and Structural Chemistry and Biological Function

Kenneth N. Raymond, Gertraud Müller and Berthold F. Matzanke

Department of Chemistry, University of California, Berkeley, California 94720, USA

Table of Contents

I Introduction

In 1911, two scientists discovered that all mycobacteria needed an "essential substance" that was vital to their growth [1]. Almost forty years later the growth factor was isolated by the crystallization of the aluminium complex [2]. A different purification procedure eventually led to the isolation of the metal free growth factor [3] and subsequently to the establishment of its principal chemical properties [4]. This compound was named mycobactin [5]. Another thirty years later we know that the mycobactins belong to the siderophores, an important class of microbial growth promoting agents. Siderophores are low molecular weight (500–1000 Daltons) ferric ion chelating agents with remarkable chemical properties which are excreted by microorganisms under conditions of iron deficiency. In early papers siderophores (from the greek: σιδεροσ = iron; φορευζ = carrier) have been called siderochromes or sideramines.

Iron is generally required for an enormous variety of metabolic processes in virtually all organisms, with the possible exception of some lactobacilli. Since ferric ion at physiological pH is very insoluble, siderophores have been evolved by microorganisms to solubilize and assimilate this critically important metal ion. However, within the last two decades it has been shown in many laboratories that the importance of siderophores is not restricted merely to supply cells with ferric ion.

In addition to their major role in iron solubilization and transport, siderophores have been identified as germination or sporulation factors within selected organisms [6]. Invading microorganisms exposed to circulating blood produce siderophores to compete for iron with the human transport protein transferrin, thus constituting one aspect of virulence and pathogenicity [7]. Pathways for iron uptake in plants are poorly understood, but it is fairly well established that root-colonizing nonpathogenic microorganisms could prevent invasion of pathogenic strains by excreting siderophores which bind iron strongly and make it unavailable for potentially harmful microorganisms [8,9]. A siderophore is employed clinically for iron removal from the body upon acute iron poisoning [10] and in the treatment of diseases like β-thalassemia [11]. In addition, siderophores have served as models for syntheses of Fe(III) and Pu(IV) complexing agents for the treatment of iron- or actinide-poisoning (chapter III) [12–14]. There are remarkable similarities between Pu(IV) and Fe(III) which explain much of the biological hazard posed by plutonium [14,15].

Not only are the physical properties of siderophores exciting from a chemical point of view, but they also yield important information on biological mechanisms involving siderophore iron complexes, e.g., the stereochemical recognition of ferric siderophore complexes by membrane receptors (chapter VII and VIII). Conclusions can be drawn from the electrochemistry of the siderophores (chapter V) about possible iron release mechanisms in microbial cells. Equilibrium thermodynamic (chapter IV) and exchange kinetics studies (chapter VI) improve the understanding of iron removal from the mammalian iron transport and storge proteins transferrin, lactoferrin and ferritin [16,17]. Moreover, sufficient thermodynamic and kinetic data will facilitate an estimation of the advantages of certain siderophores over others in their competition for iron.

Several excellent review articles on the biological [18–22], medical [10,13,23,24] and agricultural aspects [8,9,25] have been published. However, no comprehensive treatment of siderophore metal complex chemistry has appeared. That is the intention of this paper.

II Structural Features of Siderophores

In the last three decades more than 80 naturally-occurring siderophores have been isolated and characterized. Since their structural features and their biological role have been extensively and comprehensively reviewed [10,19,20,26–29] we will focus in this section only on a selection of the main representatives in order to point out the wide range of ligands which are produced by microorganisms to obtain iron from the environment (Fig. 1 and 2).

Despite the considerable structural variation found in the siderophores, their common feature is to form six-coordinate octahedral complexes with ferric ion of great thermodynamic stability. The ligating groups usually contain the oxygen atoms of hydroxamate (a) or catecholate (b) anions.

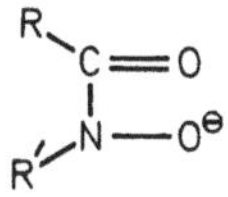

a

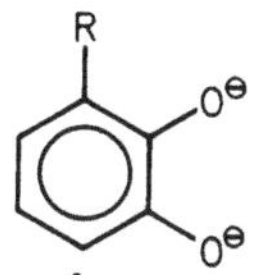

b

The ferrichromes, fusarinines, and ferrioxamines are typical trihydroxamate siderophores, while enterobactin is a cyclic tricatecholate siderophore. However, there are several exceptions which employ mixed forms of coordination. For example, aerobactin, schizokinen and arthrobactin contain, in addition to two hydroxamate groups, a α-hydroxy-carboxylate unit which completes the hexadentate ligand structure [30]. The recently isolated ferrioxamine H, with two hydroxamate and one carboxylate, is only pentadentate — the sixth coordination site presumably occupied by water [31].

Octahedral coordination by the mycobactins is achieved through two hydroxamate groups, a single phenolate group and an oxazoline nitrogen [32]. There is some evidence that agrobactin and parabactin also utilize an oxazoline nitrogen in their coordination, with the other five ligating atoms being phenolate oxygens [33]. Pseudobactin and pseudobactin A are siderophores which show the extraordinary feature of containing L- as well as D-amino acids. The hexapeptide chain is linear and consists of the following sequence: L-Lys, D-threo-β-OH-Asp, L-Ala, D-allo-Thr L-Ala, D-N^{δ}-OH-Orn. The structure contains a quinoline derivative attached through the L-Lys residue. The iron is then bound by the following functional groups: two phenolates, one hydroxamate, and a α-hydroxy-carboxylate unit [34,35].

Each of the siderophores cited above forms 1:1 complexes with iron(III). In contrast, the dihydroxamate ligand rhodotorulic acid (RA), is only tetradentate and hence has to form 2:3 complexes with iron at pH 7 [36]. Another curiosity among the siderophores is the occurrence of thioformin (N-methyl-thioformohydroxamate) which has been isolated from bacterial cultures of *Streptomyces* and *Pseudomonas* species. Thioformin acts as an antibiotic and the 3:1 complex with iron involves coordination by three oxygen and three sulfur atoms [37].

The biological rationale for the production of siderophores by microorganisms is the irreplaceable role of iron in nearly all oxidation and reduction processes of cells, combined with the extreme insolubility of ferric hydroxide at physiological pH. At pH 7 the equilibrium concentration for ferric ion (as free [Fe^{3+}]) is approximately

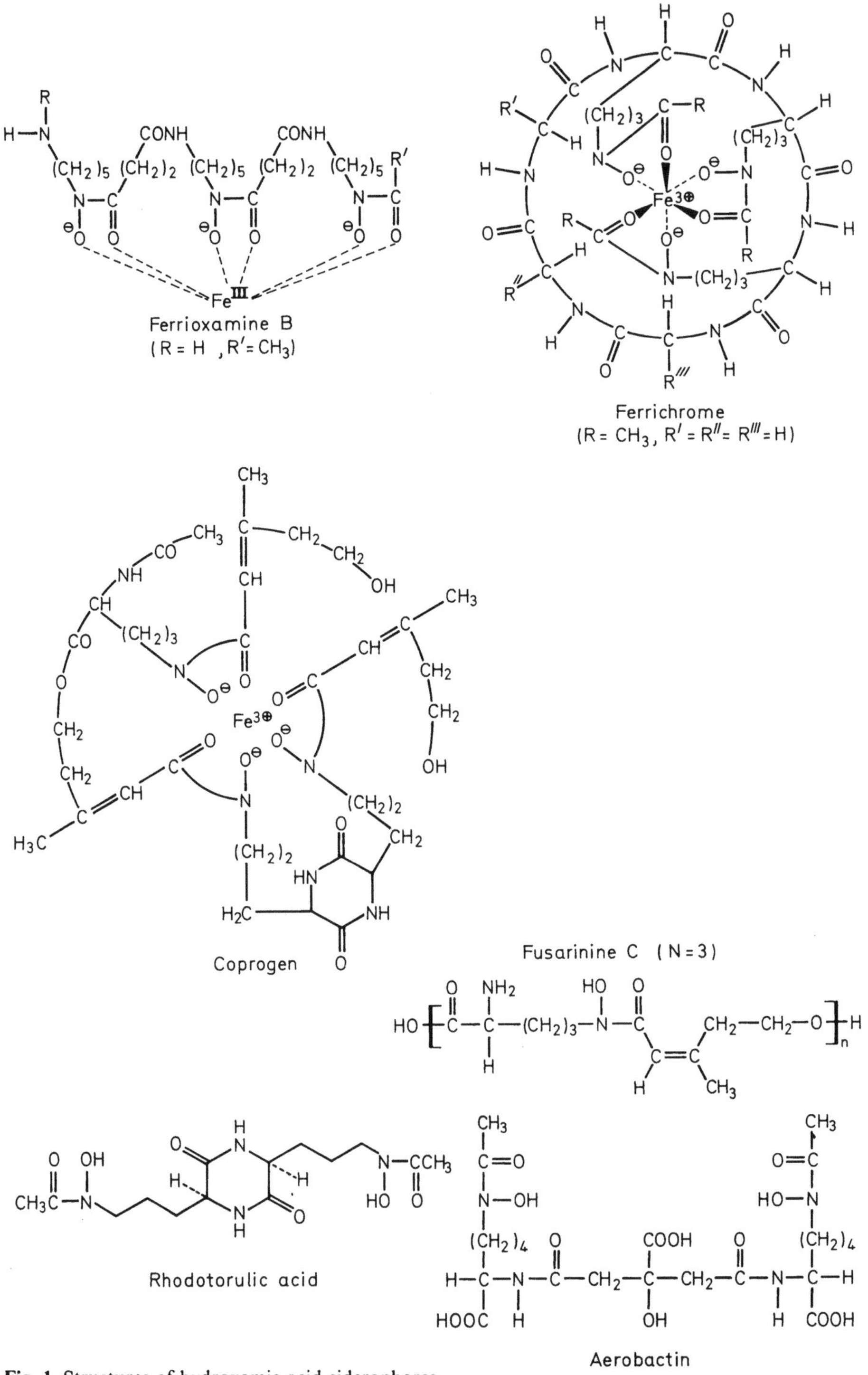

Fig. 1. Structures of hydroxamic acid siderophores

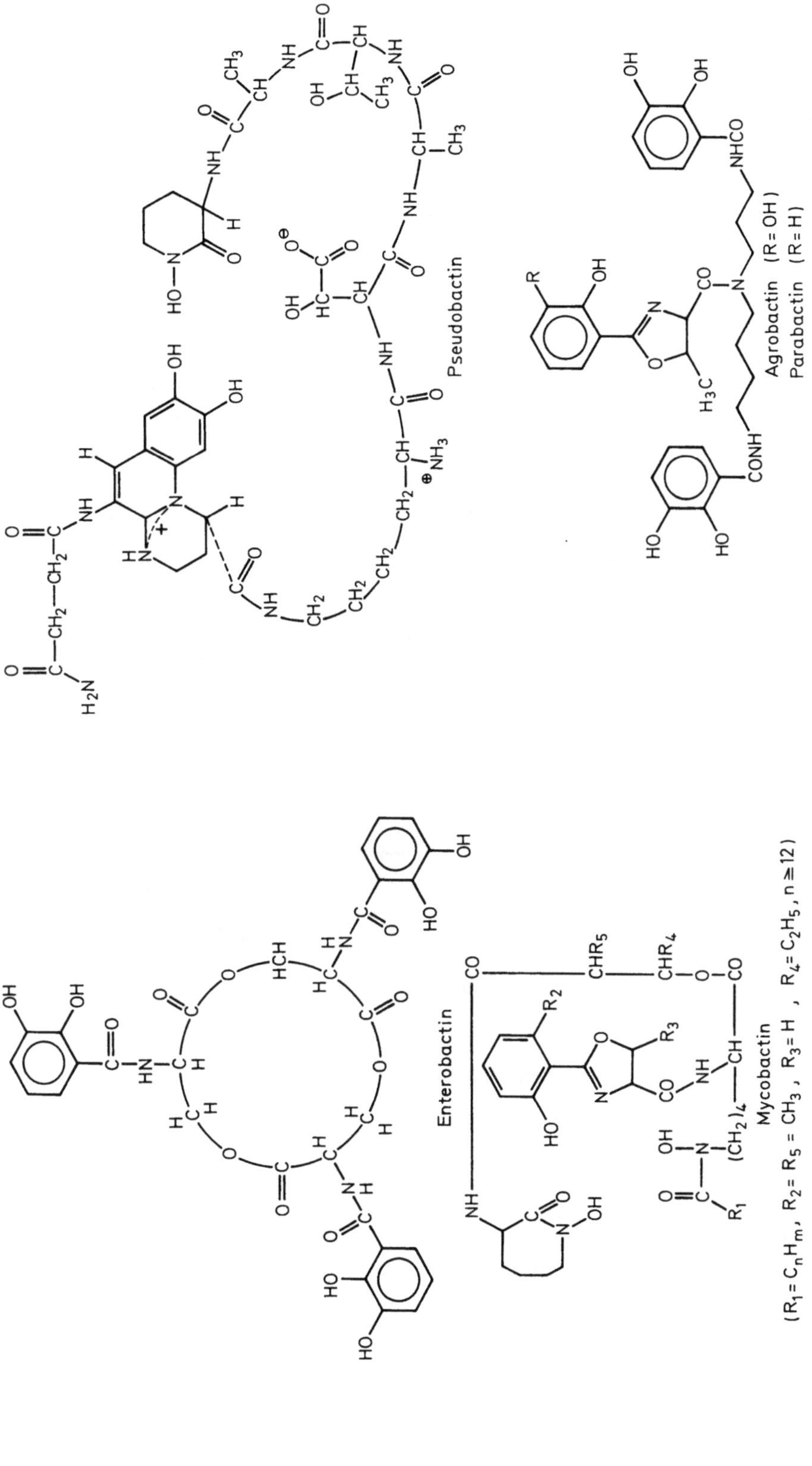

Fig. 2. Structures of catechol-containing and mixed ligand siderophores

10^{-18} M [38]. For microorganisms such as enteric bacteria, which need at least a total concentration of iron of $5 \cdot 10^{-7}$ M for optimal growth, this concentration is many orders of magnitude too low. Only powerful chelating agents such as the siderophores (see the chapter on thermodynamics for details) can mobilize iron from the environment and facilitate transport of iron into the microbial cell.

The discovery of new siderophores is continuing at a good rate. New microanalytical methods and the hypothesis that siderophores act as virulence factors of pathogenic microorganisms have focused the screening for siderophores on plant- and human-pathogenic microorganisms. Pyoverdin and ferribactin have been isolated from *Pseudomonas* species associated with plant roots, and have structural similarities to pseudobactin [39,40]. Also the N-methylphenylacetohydroxamic acid [41], malonichrome [42] and triornicin [43] have recently been characterized. Several human pathogenic microorganisms have been found to excrete siderophores: e.g. gonobactin, meningobactin [44] and pyochelins [45]. The structural formulae of the latter compounds are not yet known. A monohydroxamate from *Corynebacterium kutscheri* was found to be L-α-aspartyl-L-α-N-hydroxyaspartyl-D-cycloserine [46].

It is now known that the ability of microorganisms to respond to iron-deficient growth conditions by the excretion of iron chelating agents is not unique to this biological kingdom. Several hexa-dentate low-molecular weight iron chelating agents have been isolated as exudates from roots of water-cultured oats, wheat and rice; these are the mugineic acids [47,48], avenic acid A [49], 2′-deoxymugineic acid [50] and nicotianamine [51]. These compounds all have a similar structure which incorporates carboxylamine, hydroxyl and amine functional groups (Fig. 3a, b). However the

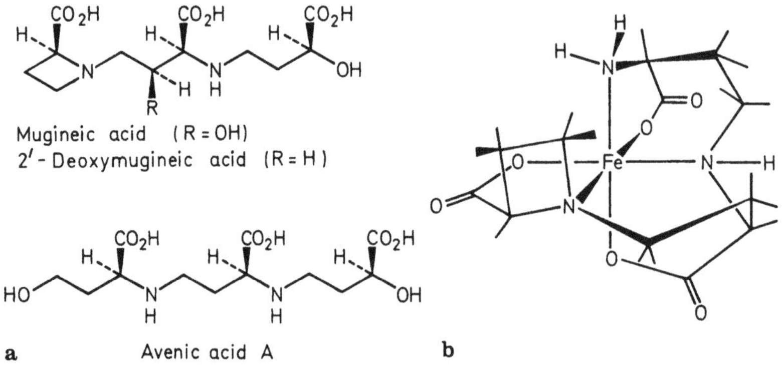

Fig. 3a. Structures of phytosiderophores; **b** dreiding model of the ferric nicotianamine complex

stability constant of mugineic acid for Fe(III) is considerably lower than that for the microbial siderophores and results in a significantly higher reduction potential than that found for the ferric siderophore complexes [$\log K_{ML} = 18.5$; E (pH 7.0) = $= -100$ mV] [48]. Whether it is legitimate to classify these plant chelating agents as phytosiderophores remains a question as long as it has not been unequivocally determined that they are iron transport agents. It is known that the addition of the ferric mugineic acid complex to iron starved rice plants results in the synthesis of chlorophyll and the disappearance of chlorophyllosis, a yellowing of the leaves [47,50].

III Synthesis of Siderophores and Their Analogs

Recently considerable effort has been devoted to the synthesis of natural siderophores, enantiomeric siderophores, and completely synthetic siderophore analogs. These compounds have been prepared as potential drugs in the treatment of iron overload, as potential antibiotics, and as probes for the stereospecific recognition and uptake of siderophore complexes by microbes. In addition, octadentate analogs have been prepared as actinide-specific sequestering agents [24].

In the treatment of β-thalassemia and certain other anemias, periodic whole blood transfusions are required. Since the iron pool of the body is almost completely recycled and since there is no specific physiological mechanism for the excretion of iron in man, continued transfusion therapy leads to a steady buildup of iron in the body [11,52], and deposition of iron in a number of organs, resulting in tissue damage and early death [53,54]. The current drug of choice for the treatment of transfusional iron overload is the trihydroxamate siderophore desferrioxamine B (Desferal®) [52,55,56]. Although Desferal is thermodynamically capable of removing iron from human transferrin, kinetically it is slow to do so. This and the short lifetime of this drug in the serum limits its effectiveness and requires large and continuous doses. Limitations of the drug include a lack of oral efficacy, moderate toxicity at the doses used, and poor effectiveness of iron removal in the absence of long-term subcutaneous infusion and the coadministration of ascorbate [57–59].

Because of the limitations of Desferal, there has been a wide search for new drugs with increased efficiency. Ideally, such agents should be orally active, resistant to degradation or metabolization in the gastrointestinal tract, the bloodstream, the liver and the kidneys [60]. Furthermore, the drug should be effective at concentrations much lower than the gram amounts per day presently used for Desferal, should show a relatively long retention time in the body, and should have no adverse side effects. The design of such drugs continues to use the siderophores as the principal model.

Since there are some naturally-occurring antibiotics such as the albomycins, which contain the structural features of the siderophores with an appended antibiotic functional unit [61,62], synthetic or derivative siderophores may be important new antibiotics [63]. In principle these are very effective, since the siderophore antibiotics are recognized as species-generated iron transport agents at the membrane receptor site of producing microorganisms [64].

Siderophore-mediated iron incorporation into microbial cells involves highly-specific recognition by receptors at the cell surfaces. In the case of enteric bacteria, these receptors consist of either one or an ensemble of proteins with a molecular weight of some ten thousand Daltons [65]. In order to study the receptor-substrate interaction, isolation of these proteins is highly desirable. However this is very difficult and has been successful only in a few cases, using genetic techniques in the case of *E. coli* [66]. In addition, since the isolated protein behavior may not reflect its behavior in vivo, chemical modification of the siderophore substrate provides an alternative. Such changes have involved changing the chirality of the molecule, systematic variation of the backbone and substituents, and alteration of the overall size of the ligand. Many of the key aspects of microbial siderophore uptake mechanisms have been revealed through time- and concentration-dependent measurements with radioactively labeled synthetic analogs (see chapter VIII).

1 Hydroxamate Siderophores and Their Analogs

The first synthesis of a siderophore was the preparation of ferrioxamine B over 20 years ago in order to confirm the chemical structure of this natural product [67]. Synthesis of the other hydroxamate containing siderophores has as a central problem preparation of the constituent ω-N-hydroxy amino acid in an optically pure form. The most important such subunit in hydroxamate siderophores is N^5-hydroxy ornithine. This is a chiral building block of the diketopiperazine-containing siderophores (rhodotorulic acid [68], dimerum acid [69], coprogen [70] and coprogen B [69]), the cyclic hexapeptides of the ferrichrome family [27], the fusarinines [71-73] and the antibiotic ferrichrome derivatives albomycines δ_1, δ_2 and ε [61,62].

In an early approach, Keller-Schierlein synthesized rhodotorulic acid, dimerum acid [74] and ferrichrome [75] by the use of ε-Nitro-L-norvaline with standard peptide synthesis techniques. The hydroxylamines are formed in the second to last step by reduction of the nitro groups (Fig. 4a). Several years later the same group employed the inverse synthetic pathway. In order to obtain enantio-ferrichrome [76], the cyclic hexapeptide was prepared by peptide synthesis, starting with optically pure D-ornithine. During oligomerization, N^5 was protected by the benzyloxycarbonyl group (CBZ). After deprotection, the N^5-amine was transformed to a Schiff base, then oxidized to an oxaziridine ring, followed by acetylation and hydrolysis to give the hydroxamic acid. In spite of the rather drastic reaction conditions, the peptide structure was not destroyed. The product, enantio-ferrichrome, has been used as a probe in stereospecific uptake in microorganisms.

Isowa et al. developed an alternative strategy: synthesis of protected N^5-hydroxy-ornithine from achiral building blocks and enzymatic resolution of the racemic product to its enantiomers [77]. The hydroxylamines were protected with tosyl- and benzyl-groups because these protecting groups are stable to the reaction conditions employed in the peptide formation and in the deprotection of the α-amino group. With this key compound, rhodotorulic acid [78], its enantiomer [78] and ferrichrome [79] were synthesized by straightforward peptide synthesis (Fig. 4a).

Recently, the preparation of aerobactin has been reported [80]. Aerobactin, as well as the mycobactins and exochelins [4,81,82], contain N^6-hydroxylysine units. The previously described synthetic procedures are based on the reduction of nitro groups, the oxidation of amino groups, or the use of protected N^5-hydroxy ornithine in order to obtain the desired hydroxamic acid. In the case of the aerobactin synthesis, O-benzyl-protected hydroxamates are directly coupled to an activated amino acid (see Fig. 4b). The alkylation agent was formed by transformation of ε-hydroxynorleucin *1* into the reactive bromide. The alkylation was accomplished by refluxing the respective bromide with O-benzylacetohydroxamate *2*. Deprotection of the N^1-amino group enabled coupling with anhydromethylenecitryl chloride *3*, followed by hydrolysis and deprotection of the remaining functionalities, yielding aerobactin. In principle the alkylation can be performed under conditions compatible with the chirality and the selective removal of α-N and α-C protecting groups, thus representing very versatile intermediates. This has been demonstrated by the same group through the syntheses of arthrobactin [83], schizokinen [83], schizokinen A [83] and mycobactin S2 [84].

In addition to the total synthesis of chiral hydroxamate siderophores, several

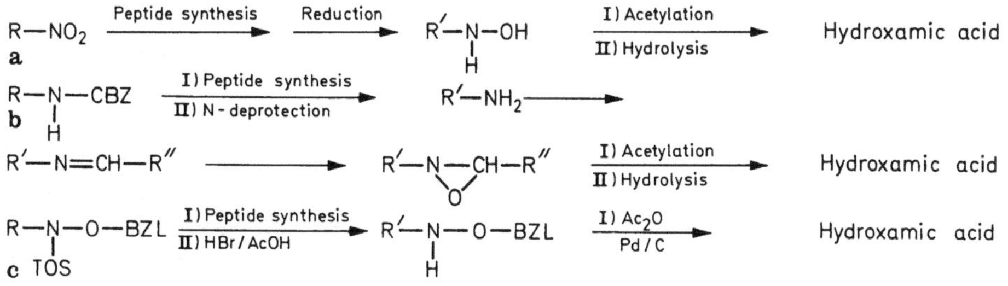

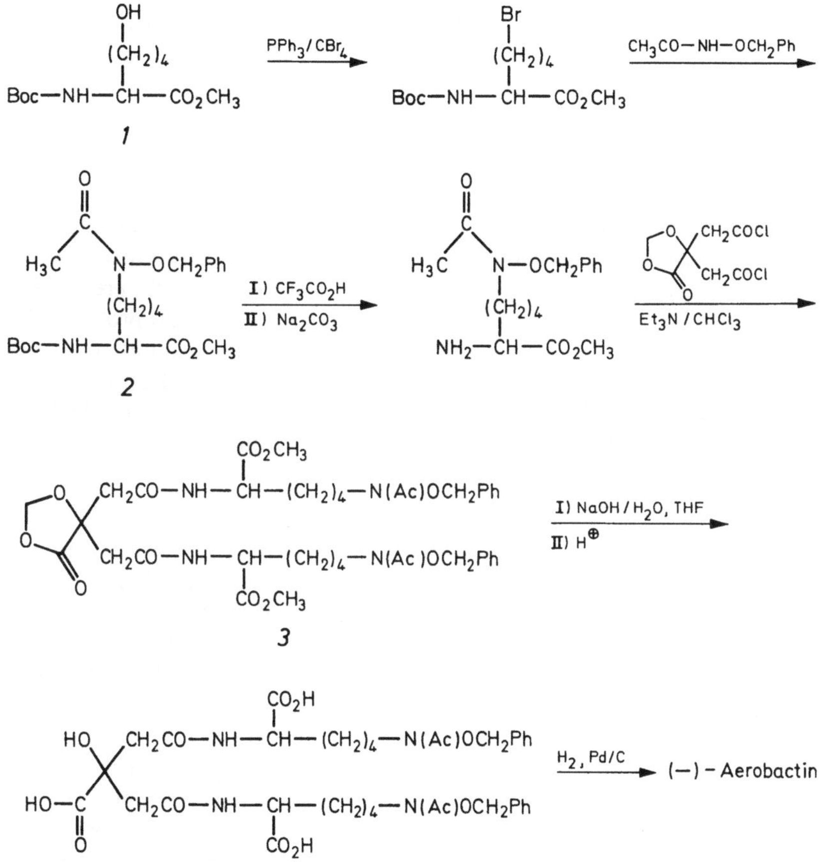

Fig. 4a. Pathways of hydroxamic acid syntheses of siderophores containing N^5-hydroxyornithine building blocks. a) Starts with ε-nitro-L-norvaline. Schemes b) and c) start with ornithine derivatives protected at N^5 with CBZ (benzyloxycarbonyl) and Tos (toluenesulfonate), BZL (benzyl), respectively
Fig. 4b. Scheme of the synthesis of aerobactin. Protected ε-hydroxynorleucine 1 (P = BOC, P′ = CH_3) is transformed into the protected hydroxamic acid 2. Deprotection of the α-NH_2 group enables alkylation of the *o*-substituted hydroxamate to give compound 3

synthetic analogs of the dihydroxamate rhodotorulic acid have been prepared. Atkins and Neilands synthesized retro-rhodotorulic acid by cyclic dimerization of L-aspartic acid methyl ester followed by the reaction with N-methylhydroxylamine [85]. The term retro indicates that the synthesized compound is rhodotorulic acid with the hydroxamate nitrogen and carbon interchanged.[1] Furthermore, a series of dimeric N-isopropylhydroxamic acids, separated by carbon chains of four to ten, have been formed from the corresponding acid chlorides by reaction with N-isopropylhydroxylamine [86]. The dimeric ferric complexes of these simple analogs, which exhibit different iron-iron distances and overall sizes, have been used as probes for the evaluation of receptor-complex interaction in iron uptake studies of *Rhodotorula pilimanae* (see chapter VIII).

Very recently a semisynthetic catechoyl derivative of ferrioxamine B has been prepared [87]. Acylation of Desferal® with four equivalents of 2,3-diacetoxybenzoyl chloride under Schotten-Baumann conditions followed by methanolic aminolysis yielded the products N-(2,3-dihydroxybenzoyl)desferrioxamine B and N-(2,3-dihydroxy-4-carboxybenzoyl)desferrioxamine B.

2 Catechol Siderophores

There are two catecholate siderophores which may be chosen as model compounds for synthesis: the cyclic enterobactin and the linear parabactin precursor N^1,N^8-bis(2,3-dihydroxybenzoyl)spermidine. Both of these natural products are capable of the rapid removal of iron from transferrin, the human iron transport protein [89,90]. The synthesis of these and other catecholate ligands routinely requires protection of the phenolic oxygens (for example, by methyl, benzyl or acetyl groups). Very few preparations of catechol-containing siderophores have appeared in which the unprotected 2,3-dihydroxybenzoyl group is used in the synthesis [91,92].

a Enterobactin and Its Analogs

Of the siderophores so far examined, by far the most powerful iron chelator is enterobactin [93] (enterochelin [94]), a cyclic triester of 2,3-dihydroxybenzoylserine that is produced by enteric bacteria such as *E. coli*, *S. typhimurium* and *K. pneumoniae*. Despite its enormous affinity for iron ($K_{ML} \simeq 10^{52}$, see Table IV), enterobactin itself is not suitable for chelation therapy because of both its chemical and biological properties. The triester backbone structure of enterobactin is very susceptible to hydrolysis, which occurs at a significant rate at physiological pH and body temperature. The free ligand also has a very low aqueous solubility [95,96]. Furthermore, the ferric complex is a powerful growth agent for enteric bacteria since it supplies the iron to these organisms and this in turn can cause toxic bacteremia, as demonstrated in iron overloaded mice [97]. However, both the chemical and biological disadvantages of enterobactin can easily be ameliorated by chemical modification and so enterobactin remains a promising model on which to base the design of new synthetic ferric sequestering agents.

---

1 It should be mentioned that a paper on the synthesis of another retro compound, retro-ferrichrome, is in press [88].

The first synthesis of enterobactin was reported in 1977 by Corey and Bhattacharya [98]. Their approach involved the formation first of the cyclic triester L-serine backbone structure and then the attachment of the 2,3-dihydroxybenzoyl groups (see Fig. 5a). In the course of the ester bond formation the amino groups were protected by benzyloxycarbonyl groups. The protecting groups for the hydroxyl and carboxyl groups were THP and the phenylacylester, respectively. Coupling of the deprotected acid and hydroxyl groups was achieved using an activated imidazolethioester. Repetition of this procedure twice gave the cyclic trimer. After deprotection of the amino groups, coupling with 2,3-dihydroxybenzoyl chloride gave enterobactin in relatively low yield.

This reaction scheme was slightly modified by Rastetter et al. in order to synthesize enterobactin and the mirror image ligand, enantio-enterobactin [99]. The latter compound was shown to be biologically inactive in supplying *E. coli* with iron. An unsuccessful attempt was made to prepare the corresponding cyclic trilactam by the same procedure; the cyclization apparently failed because intermolecular condensation proceeded faster than cyclization.

A completely different approach to the total synthesis of enterobactin has been reported by Shanzer and Libman [100]. This is based on a single-step conversion of the tritylated serine-lactone to the cyclic triester enterobactin backbone via the use of a cyclic organotin compound as template. This is followed by subsequent replacement of the trityl protecting groups by catechol residues to yield enterobactin.

The basic premise of the synthesis of synthetic analogs of the siderophores is to make structures which are functionally the same as the siderophores but which avoid the chemical or biological problems associated with them (for example enterobactin) and at the same time involve, ideally, relatively simple preparative procedures. For this purpose, mesitylene and various cyclic structures including triamines and cyclododecane derivatives have been employed as backbones. The ligands synthesized by Raymond, Weitl and co-workers have for convenience been assigned acronyms. The acronyms are based on a suffix, CAM (which indicates that these are catechoylamides), and a prefix (which indicates the type of backbone to which the catechol units have been appended, ME = mesitylene, LI = linear, CY — cyclic amine). For the cyclic and linear amine derivatives, the numerals before the ligand name indicate the number of bridging methylene units in the amine chain. When the catechol groups have been functionalized another letter is added to indicate the functional group. Thus, LICAM-S is the ligand in which each of the catechol groups has been sulfonated in the 5 position while LICAM-C is carboxylated in the 4 position (see Fig. 5b and Fig. 7a).

In order to obtain enterobactin analogs with a benzene backbone, trimesoyl chloride (Fig. 5b) was treated with ammonium hydroxide or various amines (R) to give the amide. In the case of TRIMCAM, the synthesis was completed at this step. After reduction of the amide to the amine, acid chloride R_1 was added. Finally, the catechol-O-protecting groups are removed to yield the hexadentate ligands MECAM [101], Me_3MECAM [102], TRIMCAM [103], and (NAc)MECAM [102]. It should be noted that TRIMCAM is an isomer of MECAM in which the methylene and carbonyl groups have been interchanged. This makes a major difference in their coordination chemistry. An independent synthesis of MECAM was simultaneously reported by Rastetter et al. [104]. Both MECAM and Me_3MECAM have been tested for their bioactivity in *B. subtilis* [105] and *E. coli* [104, 106]. Since each supported growth, the proposed mechanism of iron utilization from enterobactin via enzymatic ester hydrolysis comes into question (see chapter V).

Sulfonation of these ligands was designed to serve a number of purposes. It stabilizes the catechol groups against oxidation to the corresponding quinone through the electron withdrawing effect of the sulfonate groups. It also increases the otherwise very low water solubility and lowers substantially the ligand protonation constants (see chapter IV, section 1). The sulfonation of each of the catechol rings proceeds quite smoothly in high yield to give only the 5-sulfo derivatives [102, 103].

A saturated carbocyclic enterbactin analog was synthesizeed from cyclododecane *cis* 1,5,9-triol [107] and the corresponding triamine [108, 109]. The latter was obtained from the triol upon treatment with azide, followed by catalytic reduction. Formation of the triester and triamide was achieved by reaction with O-protected 2,3-dihydroxybenzoyl chloride. The triamide analog was employed in uptake studies with *E. coli*. Like MECAM and Me_3MECAM, growth promotion of siderophore auxotroph mutants was observed [110].

b Spermine and Spermidine Derivatives

In 1975 two catechol-containing linear siderophores, from *Paracoccus denitrificans*, were reported [111]; N^1,N^8-bis(2,3-dihydroxybenzoyl)spermidine and parabactin A. The former catechoylamide is a precursor of parabactin (Fig. 2), whereas the latter is a decomposition product of the same compound, generated by cleavage of the oxazoline ring into the threonyl moiety under acidic conditions.

In a very recent total synthesis of parabactin [91], N^1,N^8-(2,3-dimethoxybenzoyl)-spermidine was combined with N-protected L-threonine via the N-hydroxysuccinimide ester to form the N^4-amide *2* (see Fig. 6). Subsequently the amine protecting carbo-

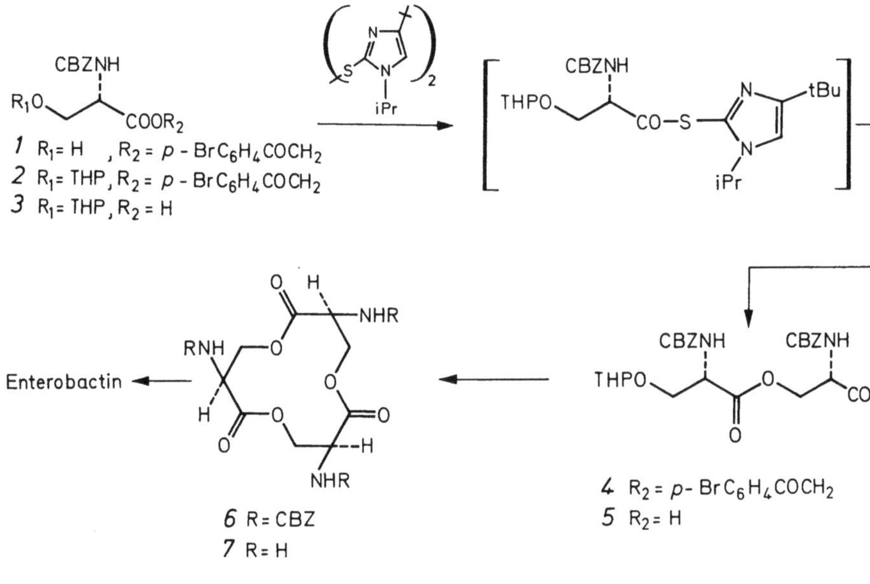

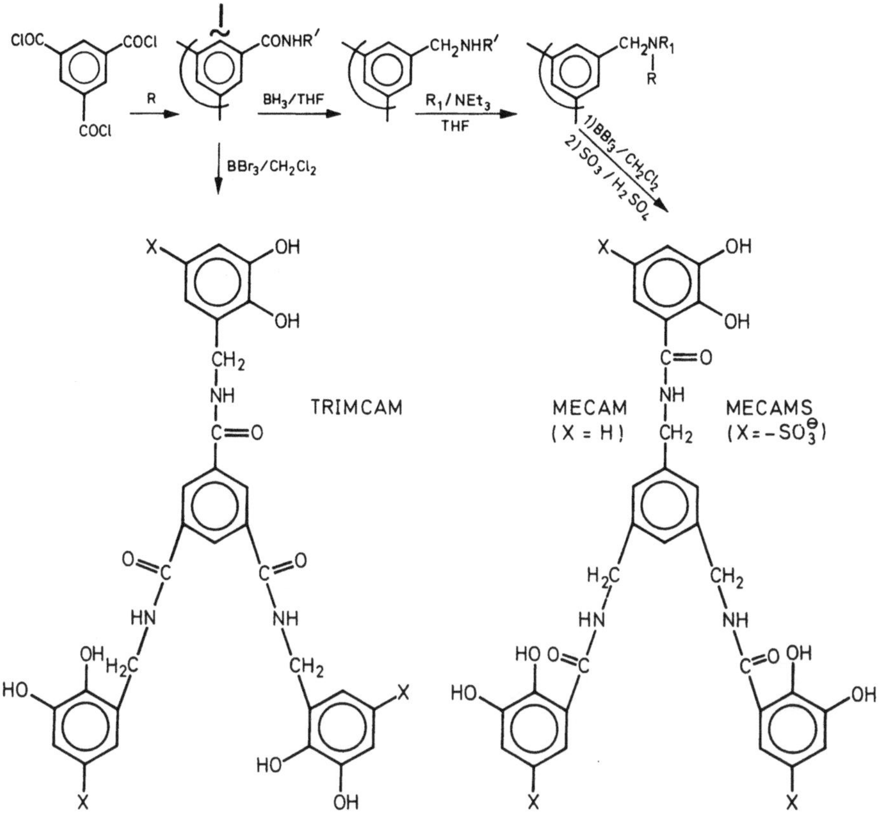

	MECAM	(Me)$_3$ MECAM	(NAc) MECAM	TRIMCAM
R	NH_4OH	CH_3NH_2	2,3 - Dimethoxybenzylamine	
R′	H	$-CH_3$	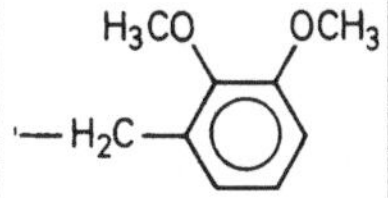	———
R$_1$		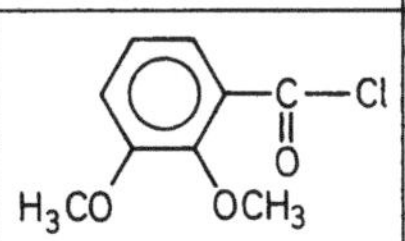	$CH_3-C(=O)-Cl$	———

Fig. 5a. A synthesis of enterobactin according to Ref. [98]
Fig. 5b. General synthetic scheme of mesitylene type enterobactin analogs. TRIMCAMS and (NaAc)-MECAMS are obtained by sulfonation of TRIMCAM and MECAM

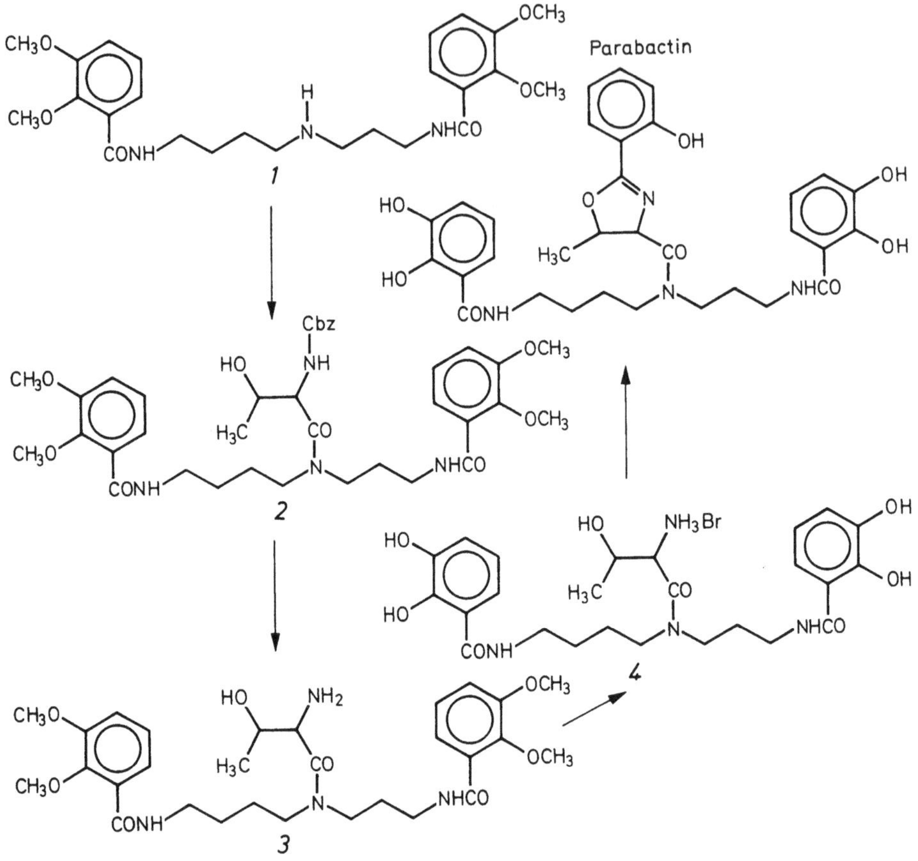

Fig. 6. Synthesis of parabactin, starting with N^1,N^8-bis(2,3-dimethoxybenzoyl)spermidine. The final step (4-parabactin), involving the formation of the oxazoline, is performed without catechol protecting groups

benzoxy group and the catechol protecting methyl groups were removed. The final and most critical step in the procedure involved the condensation with 2-hydroxy-benzimino ethylether in order to form the oxazoline. In order to confirm the structure of the closely related molecule agrobactin, the synthesis of the decomposition product, agrobactin A, was carried out [112].

As direct analogs of the threonine conjugate of spermidine, a multitude of trimeric

Fig. 7a. General syntheses of catechol siderophore analogs containing a linear cyclic amine backbone. ▶ The amine may vary in the chain length (for spermidine m = 3, n = 4, $R_1 = R_2 = H$) and in the substitutents of the terminal amines [E.g. alkylated according to Ref. 117; spermine: m = 3, n = 4, $R_1 = H$, $R_2 = -(CH_2)_3-NH_2$]; **b** Synthesis of a cyclic secondary amine backbone; **c** Synthesis of methyl 2,3-dimethoxyterephthaloyl chloride. This compound is used as coupling agent in the amide formation of Figure 7a in order to obtain catechols carboxylated in the 4-position, e.g., LICAMC

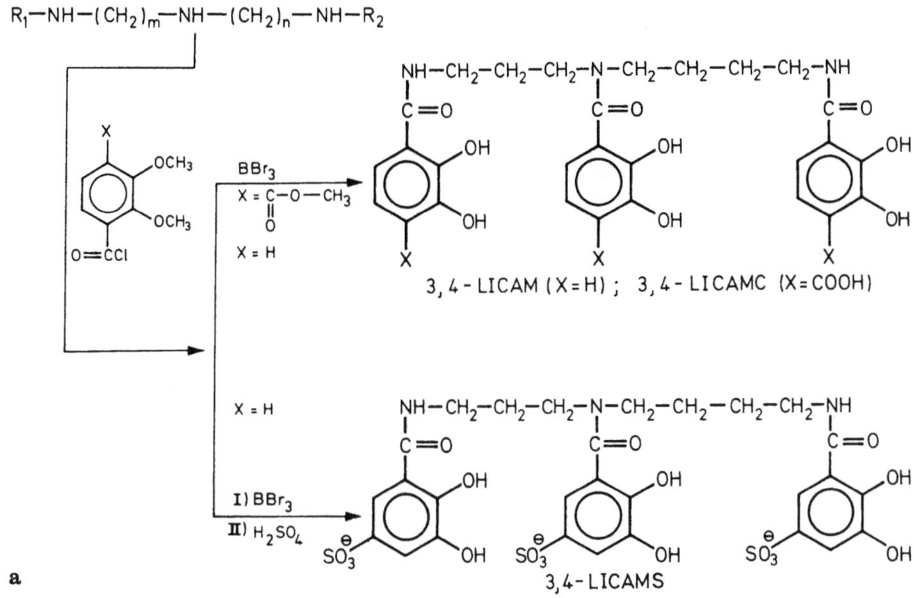
R1—NH—(CH2)m—NH—(CH2)n—NH—R2
X
OCH3
OCH3
O=CCl
BBr3
X = C(=O)—O—CH3
X = H
NH—CH2—CH2—CH2—N—CH2—CH2—CH2—CH2—NH
C=O
OH
OH
X
3,4-LICAM (X=H); 3,4-LICAMC (X=COOH)
I) BBr3
II) H2SO4
SO3
3,4-LICAMS

a

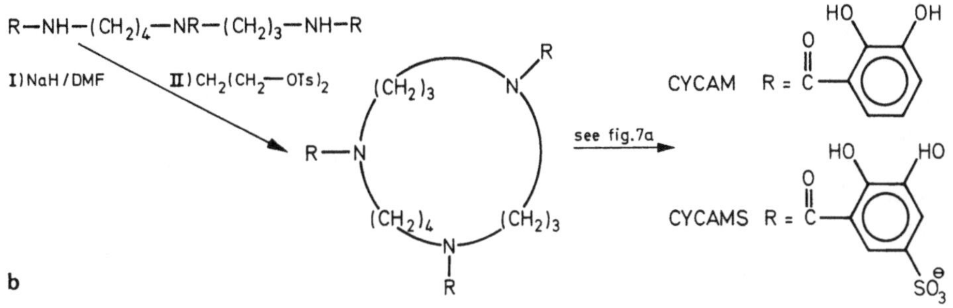
R—NH—(CH2)4—NR—(CH2)3—NH—R
I) NaH/DMF
II) CH2(CH2—OTs)2
(CH2)3
(CH2)4
R—N
see fig.7a
CYCAM
CYCAMS
HO
OH
SO3

b

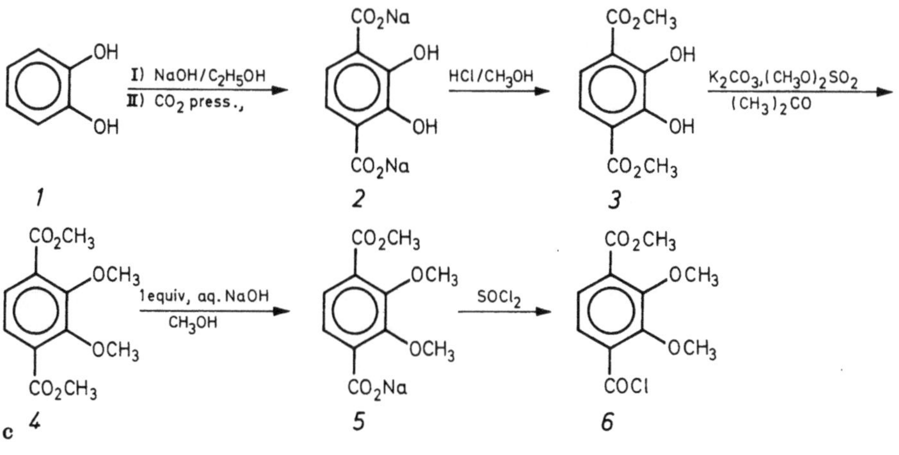
I) NaOH/C2H5OH
II) CO2 press.,
CO2Na
HCl/CH3OH
CO2CH3
K2CO3,(CH3O)2SO2
(CH3)2CO
1equiv, aq. NaOH
CH3OH
SOCl2
COCl
1
2
3
4
5
6

c

and tetrameric 2,3-dihydroxybenzoyl conjugates have been designed which incorporate certain linear and cyclic amines. The general reaction is shown in Fig. 7a. The reaction of an oligoamine with a substituted, O-protected benzoylchloride yields the corresponding amides. In preceding steps the terminal alkylated or cyclic amines (Fig. 7b) are combined with 2,3-dimethoxybenzoyl chloride and methyl 2,3-dimethoxyterephthaloyl chloride (Fig. 7c).

Cyclization of the amines is achieved by N-protection with tosylate and formation of the terminal dianion *in situ* under treatment with sodium hydride, followed by reaction with α,ω-ditosylalkane [92] (see Fig. 7b). Deprotection in concentrated H_2SO_4 yields the cyclic free amine. Finally, by reaction of the amine with the described acid chloride and deprotection of the catechol oxygens, 3,4,3-CYCAM is obtained. In thermodynamics studies it was determined that CYCAM ligands are slightly less effective chelating agents than the corresponding linear or mesitylene derivatives (see chapter IV) [113].

A most important property of any sequestering agent to be used in chelation therapy is oral activity. One of the few agents to show any oral efficacy as an iron removal drug is 2,3-dihydroxybenzoic acid [114]. Therefore, the 2,3-dihydroxyterephthalate conjugates of spermine and spermidine have been synthesized [115] according to the general scheme outlined above. While the neutral catechol derivative is insoluble in water, the carboxylate derivative shows increased water solubility and lower toxicity.

The efficacy of an orally administered drug will be dependent on the extent to which it is absorbed in the gastrointestinal tract. Passage through the gastrointestinal epithelium is, for most drugs, a passive diffusion process which is more facile for more lipopholic drugs [116]. Since the coordination chemistry of In(III) and, especially, Ga(III) are very similar to that of Fe(III), the new chelating agents may be of additional use in radiopharmeceuticals incorporating ^{111}In and ^{67}Ga. In order to achieve desired tissue distribution and imaging *in vivo*, a much greater lipophilicity of the ligand may be of critical importance. This was considered in the design of spermine and spermidine derivatives by selective reductive alkylation of the terminal nitrogen with acetone or aldehyde under hydrogen [117].

Of the actinides, plutonium is potentially a particularly dangerous biological hazard because of the chemical and biological similarities between Pu(IV) and Fe(III). With this similarity in mind, a series of tetracatechol ligands with both linear and cyclic tetramines as backbone were synthesized in accordance with the general scheme outlined above [12, 24, 118] (see Fig. 7a and b). The eight-coordinate nature of Pu(IV) is satisfied via four catecholate moieties. In the case of the more effective linear octadentate complexing agents, the size of the central cavity formed by the four appendant catechol groups may be adjusted by changing the number of bridging methylene units in the tetramine platform. The length of the linear backbone has been determined from the structure of simple tetrakis catecholato actinide(IV) compounds [119] to allow for the best fit of the large Pu(IV) ion. The optimum geometry appears to be achieved when the central bridge is butylene and the terminal bridges are propylene; this is the natural product spermine. The corresponding synthetic catechol is 3,4,3-LICAM, which in animal tests has been shown to bind Pu(IV) effectively [120], although full encapsulation of the metal ion only occurs above physiological pH [121].

IV Equilibrium Thermodynamics of Siderophore Iron Binding

The formation constants of siderophore complexes define the thermodynamic limits for the conditions in which siderophores can compete for iron and extract it from a weaker substrate; thus eventually providing superior propagation conditions for that organism from which the superior complexing agent was excreted. In addition, the tenacity of iron binding may determine the mechanism of iron metabolization by means of ligand exchange, reductive removal or ligand destruction. Equilibrium formation constants are also important for considerations dealing with the synthesis of ferric ion sequestering agents. Here the formation constants can be used as a primary criterion for the rational design of stable and metal-ion-selective ligands for use in iron chelation therapy.

1 Ligand Protonation Constants

The knowledge of ligand protonation constants is imperative for the calculation of both stability constants and the effective stability of a metal ion complex in pH regions where the ligand is at least partially protonated. In general, the greater the basicity of a chelator, the greater its affinity for Fe(III). However, when the pK_a of the ligand is substantially greater than the physiological pH, proton competition will considerably decrease the concentration of the basic form of the ligand, thus reducing the metal iron binding. Hydroxamate siderophores are fairly weak acids with pK_a's in the range of 8 to 9 (see Table I). The protons of desferricoprogen are more acidic compared

Table I. Ligand protonation constants of hydroxamic acids

	pK_1 [a]	pK_2 [a]	pK_3 [a]	pK_n	Ref.
Ferricrocin	9.92	9.01	8.14	—	214)
Coprogen	9.16	8.86	7.63	—	214)
Ferrioxamine	9.70	9.03	8.39	—	125)
Ferrichrome	9.83	9.00	8.11	—	125)
Rhodotorulic acid	9.44	8.49	—	—	36)
Acetohydroxamatic acid	9.35	—	—	—	215)
Aerobactin	9.44	8.93	4.31	3.48 3.11	36)

[a] $K_1 = [HL^{2-}]/([H^+][L^{3-}])$; $K_2 = [H_2L^-]/([H^+][HL^{2-}])$; $K_3 = [H_3L]/[(H^+][H_2L^-])$.

Table II. Ligand protonation constants of monomeric catechols

	Catechol	Sulfocatechol	DMB[a]	DMBS[b]
K_1^H	13	12.16	12.1	11.5
K_2^H	9.2	8.3	8.42	7.26
Ref.	216	216	122	102

[a] DMB = N,N-dimethyl-2,3-dihydroxybenzamide.
[b] DMBS = N,N-dimethyl-2,3-dihydroxy-5-sulfobenzamide.

Tabelle III. Ligand protonation constants and complex protonation constants of tricatecholate ligands

	log K_4^H	log K_5^H	log K_6^H	log $K_{4-6}^{av.}$	log K_{MHL}	log K_{MH_2L}	log K_{MH_3L}	Ref.
Enterobactin	9.2	8.4	7.6	8.4	4.89	3.15	—	102, 122)
MECAM	—	—	—	—	7.08[a]	5.6[a]	—	113)
3,3,4-CYCAM	9.26	8.65	7.86	8.59	9.0[b]	7.6[b]	5.9[b]	113)
MECAMS	7.26	6.44	5.88	6.5	5.74[b]	4.10[b]	3.46[b]	123)
					5.19[a]			
TRIMCAMS	8.72	8.16	7.53	8.1	—	13.66[a]	—	123)
						13.1[b]		
3,3,4-CYCAMS	8.56	7.8	7.01	7.8	6.78[a]	5.74[a]	5.54·	123)
3,4-LICAMS	8.28	7.07	6.11	7.2	5.85[a]	5.32[a]	3.05[a]	123)
(NaAc)MECAMS	9.3	8.4	7.7	8.47	—	14.48[a]	—	102)
(Me_3)MECAMS	8.52	7.57	6.72	7.6	6.61[b]	5.80[b]	—	102)

[a] Determined from spectrophotometric titration. [b] Determined from potentiometric titration. Description of acronyms of the synthetic analogs is given in Chapter III.2. For discussion of the constants see text.

with those of other hydroxamates. This is no doubt related to the aliphatic side chain, which contains a double bond conjugated to the hydroxamate group.

Catechol itself is a very weak diprotic acid with widely separated protonation constants for the two phenolic oxygens (Table II). Because of this large difference in the intrinsic acidity of catecholate oxygens, the six dissociable protons of the tri-catecholate ligands clearly divide into groups of three. The three less acidic protons (K_{1-3}) show no appreciable dissociation below pH 11 (thus being outside the range of the glass electrodes used in potentiometric titration). Since the average of the more acidic ligand protonation constants K^H_{4-6} agrees with the corresponding bidentate ligands K_2 (DMB [122] for unsulfonated, DMBS [123] for sulfonated compounds) (see Tables II and III), the extrapolation is legitimate that this will hold for the constants K^H_{1-3} as well. In addition, when the protonation constants of the sulfonated and the unsulfonated homologs are compared, it is obvious that the inductive effect of the sulfonate group lowers the constant of aromatic hydroxyl groups by about 1 log unit. In contrast, the protonation constants of TRIMCAMS and the closely related (NAc)MECAMS (Fig. 5b) are larger than those of the other ligands, since there is no electron withdrawing capacity of a carbonyl group α to the dihydroxybenzene rings. On the basis of acidity differences between catechol and DMB, the carbonyl groups appear to lower the catechol protonation constant by 0.8 log units. Using these correlations the values of $K^{H\,av.}_{1-3}$ can be estimated for each tricatechol ligand employed in these studies.

2 Complex Formation Constants

The overall ferric ion complex formation constants of siderophores cannot be determined directly at neutral pH, because the strong iron binding pulls the equilibrium (Eq. 1) to the right, exhibiting (in the case of hydroxamates) no appreciable dissociation into free ligand and free iron above pH 2.

$$Fe(III) + L \rightleftarrows Fe(III)L \tag{1}$$

One method of circumventing this problem is the spectrophotometric measurement of competition (Eq. 2) for the metal by another ligand, typically EDTA.

$$FeL + EDTA^{4-} + xH^+ \rightleftarrows Fe(EDTA)^- + H_xL \tag{2}$$

In order to convert the resultant proton dependent equilibrium constant (Eq. 3) into the conventional formation constant (Eq. 4) it is necessary to know the ligand protonation constants.

$$K_x^* = \frac{[FeL]\,[H^+]^x}{[H_xL]\,[Fe^{3+}]} \tag{3}$$

$$K_{ML} = \frac{[FeL]}{[Fe(III)]\,[L]} \tag{4}$$

Table IV. Stability constants and redox potentials of ferric natural siderophore complexes[a]

	$\log \beta_{110}$ [b]	pM[c]	log K* [b]	$\log_{MHL}$ [b]	$\varepsilon_{1/2}$, mV vs NHE	Ref.
Ferrioxamine E	32.5	27.7				216)
Coprogen	30.2	27.5	4.6	0.5	—447	214)
Ferrioxamine B	30.5	26.6	3.4	1.0	—468	125, 217)
Ferricrocin	30.4	26.5	3.3	0.53	—412	214)
Ferrichrysin	30.3	25.8	2.76			125)
Ferrichrome A	32.0				—448	218, 217)
Ferrichrome	29.1	25.2	2.13	1.49		125, 214)
Aerobactin	22.5	23.3			—336	30)
Rhodotorulic acid	(31.2)	21.9			—359	36)
Enterobactin	52	35.5	9.7	4.89	—750[d]	124, 217)
2,3-dihydroxy-benzoylserine					—350	95)
Parabactin					—673[d]	143)
Parabactin A					—400	143)

[a] All solutions are aqueous. [b] $\beta_{110} = [FeL^{3-4}]/[Fe^{3+}][L^{n-}]$, $K^* = [FeL^{3-n}][H^+]^3/[Fe^{3+}][H_3L^{3-n}]$, $K_{MHL} = [FeHL^{4-n}]/[H^+][FeL^{3-n}]$. [c] $pM = -\log[Fe(H_2O)_6^{3+}]$ when $[Fe]_T = 10^{-6}$ M, $[L] = 10^{-5}$ M, pH 7.4. [d] Measurements performed at pH 10 and estimated for pH 7. For low pH values (~ pH 4) a dramatic increase of the redox potential can be calculated (see Chapter V).

Table V. Stability constants of synthetic analogs

	pM[a]	$\log \beta_{110}$[b]	log K* [b]	λ_{max}		Ref.
MECAM	29.4	46	9.5	492 (ε = 4700)	pH 8.5	113)
				515 (ε = 3900)	pH ~ 7.0	
3,3,4-CYCAM	23.0	40	3.4	480 (ε = 4300)	pH > 9	123)
MECAMS	29.1	41	6.57	485 (ε ~ 5800)[c]	pH ~ 7.8	123)
				560	pH ~ 2.5	
(Me$_3$)MECAMS	26.9	40.6	5.21	487 (ε = 5390)	pH > 9.0	102)
3,4-LICAMS	28.5	41	6.4			123)
TRIMCAMS	25.1	41	4.43	480 (ε = 5300)	pH > 8	123)
				560	pH ~ 6	
(NAc)MECAMS	25	40.3	4.0	487 (ε = 5400)	pH > 8	102)
Enterobactin	35.5	52	9.7	495 (ε = 5600)	pH 7.3	122, 218)
DMB	(15)[d]	(43.9)		488 (ε = 4910)	pH > 8.6	122, 124)
				570 (ε = 3750)	~ pH 6.0	
DMBS	19.2			480 (ε ~ 5800)[c]	pH > 9	123)
				563	pH < 5.5	

[a] All solutions are aqueous. $pM = -\log[Fe(H_2O)_6^{3+}]$ when $[Fe]_T = 10^{-6}$ M, $[L] = 10^{-5}$ M, pH 7.4. [b] $\beta_{110} = [FeL^{3-n}]/([Fe^{3+}][L^{n-}]$, $K^* = [FeL^{3-n}][H^+]^3/([Fe^{3+}][H_3L^{3-n}]$. [c] Extinction coefficients are from figures in Ref. 123. [d] pM is below lower limit set by the K_{SP} of ferric hydroxide indicating precipitation of $Fe(OH)_3$ under these conditions.

Although the conventional form for tabulation, the formation constants are not meaningful alone in judging the relative ability of ligands to compete with one another at a given pH. This is due to differences in ligand protonation constants (the number of them and their absolute value), which define the amount of free, uncomplexed ligand [L] in aqueous solution. In addition, the protonation constants of tricatechol ligands are only estimates. In order to have a more direct ranking of the ligands under specified conditions, so called pM values have been used. Normally pM is defined as $-\log[M(H_2O)_n]^{m+}$, calculated from the observed proton-dependent formation constant K* for physiological conditions, i.e.; pH 7.4, ligand concentration 10 μm and metal concentration of 1 μM. The larger the pM the more stable the complex is under the prescribed conditions (Table IV, V). Any pM values below the limit set by the K_{sp} of $Fe(OH)_3$ indicate precipitation of ferric hydroxide under these conditions.

A comparison of the stability constants of the naturally-occurring siderophores shows a difference of twenty orders of magnitude between enterobactin ($K \sim 10^{52}$) [124] and the most stable hydroxamate complex ferrioxamine E [125]. Using the more comparable pM values, enterobactin remains still eight orders of magnitude more effective than ferrioxamine E. Enterobactin has the highest affinity for ferric ion of any iron chelator tested so far, and this has engendered the synthesis of enterobactin analogs.

The pM values of the synthetic catecholate ligands 3,4-LICAMS, 3,4,3-LICAMS, MECAM and MECAMS (see Table V) arc all exceptionally high, ranging from 28.5 to 31.5. Although these ligands are not as powerful as enterobactin, when compared with the hydroxamates (particularly ferrioxamine B, the currently used drug in iron chelation therapy), the synthetic catechols are up to 10,000 times more effective at sequestering Fe(III) at pH 7.4 and show a similar specificity for Fe(III) over other divalent metal ions [126]. These catecholate sequestering agents are both thermodynamically [102,122,123] and kinetically [16,89,102,115] able to remove ferric ion from transferrin.

There is a surprisingly large increase in the proton-dependent formation constant (log K*) value of MECAM(S) compared to TRICAM(S), especially considering that the higher ligand protonation constants of TRIMCAM(S) would tend to increase its log K*. There is only an isomeric structural variation between TRIMCAM(S) and MECAM(S), as shown in Fig. 5b. Thus the overall formation constants appear to be very sensitive to changes in the structure of the ligand. It is suspected that the exceptionally high stability constant of enterobactin is due in large part to the conformational flexibility of the triester ring, as compared to the rigidly planar benzenoid

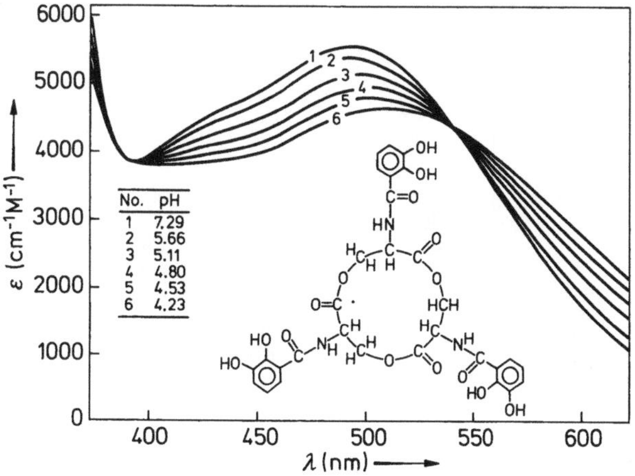

Fig. 8. The pH dependent visible spectra of ferric enterobactin

platform in MECAM(S) and TRIMCAM(S). It may even be that the triester group is predisposed toward a conformation favourable to the optimal arrangement of the coordinating groups about the ferric ion. The relatively low stability constants for both CYCAMS and CYCAM are indicative of considerable strain associated with the complete encapsulation of Fe^{3+}. This presumably results from having the amide nitrogens contained within the thirteen membered ring. In both enterobactin and MECAM, the nitrogens are exocyclic, which results in larger bridging groups between the exterior 2,3-dihydroxybenzoyl rings and the central triester ring of enterobactin or the benzene ring of the mesitylene analogs.

3 Protonation of the Ferric Siderophore Complexes

a Solution Protonation Equilibria

The stoichiometries of the principal species in solution can be determined from both the (concentration-dependent) potentiometric as well as spectrophotometric data. If there are only two siderophore species in equilibrium, spectrophotometric titration data typically show a shift of the ligand-to-ferric ion charge transfer band to longer wavelength with decreasing pH, and an isosbestic point. Using the Schwarzenbach equation (5) [125] the data should fit to a linear correlation of A_{obs} vs $(A_0 - A_{obs})/[H^+]^n$,

$$A_{obs} = \frac{A_o - A_{obs}}{K_{MH_nL}[H^+]^n} + \varepsilon_{MH_nL}C_T \tag{5}$$

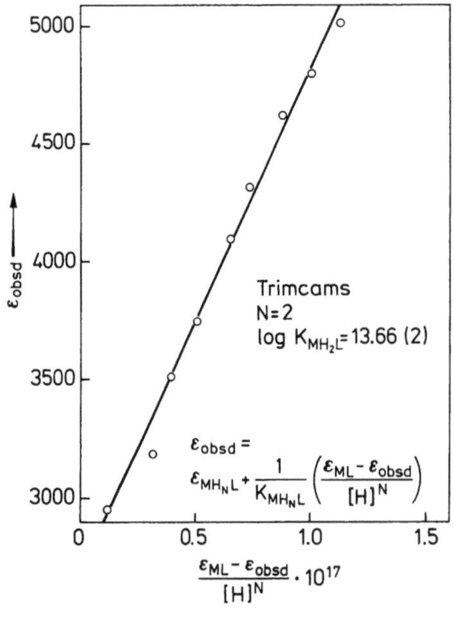

Fig. 9. Plot of $(\varepsilon_{ML} - \varepsilon_{obsd})/[H]^n$ vs. ε_{obsd} for ferric TRIMCAMS using $n = 2$, where ε_{ML} is the molar extinction coefficient of $[FeTRIMCAMS]^{6-}$ at 480 nm and ε_{obsd} is the apparent extinction coefficient at any pH (ε_{obsd} = absorbance$_{480}$/total iron). Data cover the pH range 7.32 — 6.36

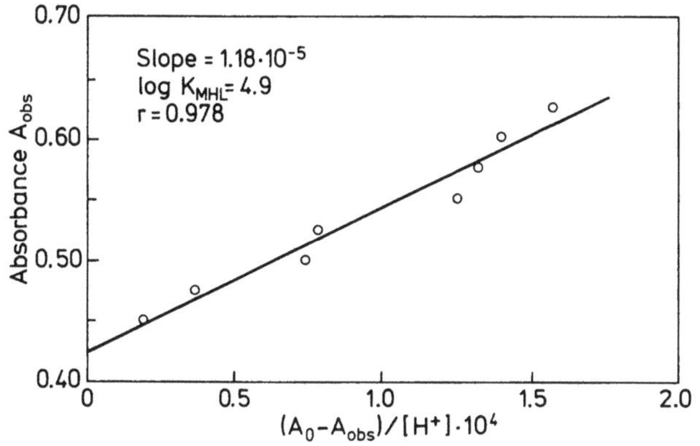

Fig. 10. A plot of the absorbance at 450 nm of ferric enterobactin solutions in 50% aqueous methanol as the pH is changed. The data are from Figure 10 of Ref. 148. The vertical axis is absorbance (A_{obs}). The horizontal axis is the function $(A_0 - A_{obs})/[H^+]$ and has been multiplied by 10^4. A linear relationship implies a single-proton stoichiometry in the reaction. A least-squares fit (the line shown) gives a corelation coefficient of 0.978, a slope of 1.18×10^{-5} (the inverse of the protonation constant) and an intercept of .425 (A_∞)

where A_{obs} is the absorbance at λ_{max} of the unprotonated species at each pH, C_T is the total concentration of siderophore, ε_{MH_nL} is the extinction coefficient of the protonated species, and A_0 is the initial absorbance at λ_{max}. The exponent n is the stoichiometric coefficient of hydrogen ion in the protonation reaction of the ferric complex. A linear plot is only obtained by choosing the appropriate value of n, which yields K_{MH_nL} from the slope (see Fig. 9 and 10).

The protonation constants K_{MHL} of ferric trihydroxamate siderophore complexes are listed in Table IV. The K_{MHL} values are in the range of 1, indicating that these complexes are stable even under very acidic conditions. Recently a complete set of potentiometrically determined protonation constants of ferric desferrioxamine B (DFOB) have been reported with the following values:[127] $K_{MHL} = 2.64$, $\log K_{MH_2L} = 0.77$ and $\log K_{MH_3L} = -1.16$. Under the very high acidic conditions employed, these authors observed the formation of a biferric complex $Fe_2(DFOH)^{4+}$, which was isolated and characterized. Using spectral and kinetic techniques, another group[128] postulated five intermediate species for the stepwise dissociation of Fe(III) from DFOB in an acidic aqueous solution (pH 1.5–pH 0). However. no biferric desferrioxamine complexes have been considered in this study.

b Catecholate Complexes — "Salicylate" Mode of Bonding

At high pH, the tricatecholate ligands derived from 2,3-dihydroxybenzoic acid bind iron via the six phenolic oxygens of the catechol rings just as do simple mono-catechols. However, the dissociation dehavior for tris(mono-catechol) complexes as the pH is lowered is quite different from that observed for the multidentate ligands which incorporate DHB groups. For the mono-catechol complexes, a simple, stepwise dissociation of individual catechol rings is observed and this is associated with the association of two protons per catechol. In contrast, the spectrophotometric behavior

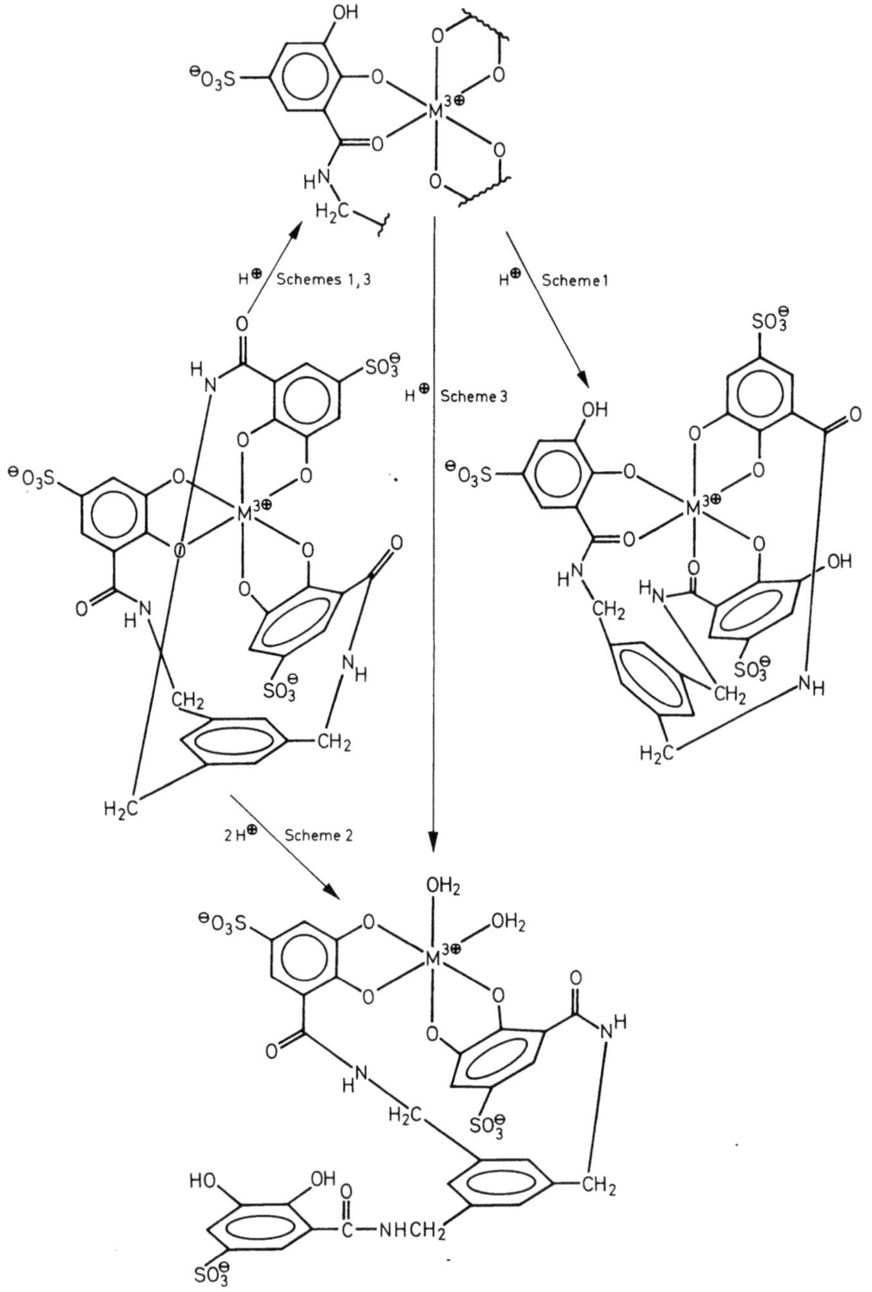

OH
⊖O3S
M3⊕
HN
H2C
H⊕
Schemes 1, 3
H⊕
Scheme 1
H⊕
Scheme 3
SO3⊖
2H⊕
Scheme 2
OH2
CH2
NH
HO
C—NHCH2

(Schwarzenbach plots)[125] of LICAMS, MECAMS[123], MECAM[113] and enterobactin[122] (Fig. 10) all show *single-proton* stoichiometries (Fig. 11, Scheme 1). These compounds all possess an amide carbonyl function which is capable of forming a bidentate chelating group with the ortho phenolate oxygen in a manner similar to that of the salicylate anion. Raymond and co-workers have proposed this mode of coordination for the protonated species of ferric catechoyl amides that display this single-proton behavior. By analogy with the coordination of salicylate anion, this has been called the "salicylate" mode of coordination. Ligands such as TRIMCAMS[123] (Fig. 9) and (NAc)MECAMS[102], which do not have the required ortho carbonyl group, do not display the "salicylate" mode of bonding for the iron complexes but rather undergo simple, two-proton reactions analogous to the simple monocatecholate complexes (Fig. 11, Scheme 2). Several kinds of physical studies have been used to support this characterization of the low pH species. Mössbauer spectra of enterobactin[129] show the creation of a fast-relaxing Fe(III), high-spin species as the pH is lowered (see Fig. 16). This indicates a change in the electronic environment of the ferric iron and is consistent with the formation of a salicylate Fe(III) complex. Direct involvement of the α-carbonyl group in coordination to the iron has been probed using Fourier transform infrared spectroscopy for both enterobactin (see Fig. 13) and some synthetic analogs. The carbonyl stretching frequency (in the region 1593–1640 cm^{-1}) is affected by both ligand deprotonation and metal iron coordination (see Table VI). The examination of a bis-catecholate complex of cupric MECAMS

Table VI. Comparison of the amide I stretching frequencies for catechoylamide and salicylamide complexes. Spectra were taken in KBr

		$\nu_{(C=O)}$
Ligand	Enterobactin	1640
	MECAM	1635
Catecholate complexes	$[Fe(ent)]^{3-}$	1593
	$[Fe(MECAM)]^{3-}$	1605
Salicylate complexes	$[Fe(H_3ent)]^0$	1640, 1620
	$[Fe(H_3MECAM)]^0$	~1610

showed that a single, uncoordinated ligand arm can be detected even when the other two arms are coordinated to the metal[130]. Furthermore this study showed that all three catechoyl arms of ferric MECAMS remain associated with the metal ion when protonated (see Fig. 12).

With the characterization of the salicylate mode of binding with increasing acidity, the pH dependent equilibrium behavior of enterobactin (and similarly its direct

◀ **Fig. 11.** Possible protonation schemes of tris catecholate metal complexes. The compound shown is $[M^{3+}(MECAMS)]^{3-}$. In scheme 1 the metal complex undergoes a series of two overlapping one-proton steps to generate a mixed salicylate-catecholate coordination about the metal ion. Further protonation results in the precipitation of a tris salicylate complex (e.g. enterobactin, MECAM). This differs from scheme 2, in which a single two-proton step dissociated one arm of the ligand to form a bis(catecholate) chelate (e.g. TRIMCAM). Scheme 3 incorporates features of scheme 1 and 2. In this model the metal again undergoes a series of two overlapping one-proton reactions. However, unlike the case of scheme 1, the second proton displaces a catecholate arm, which results in a bis-(catecholato) metal complex. This scheme is discussed for Ga and In complexes of enterobactin analogs in Ref.[141]

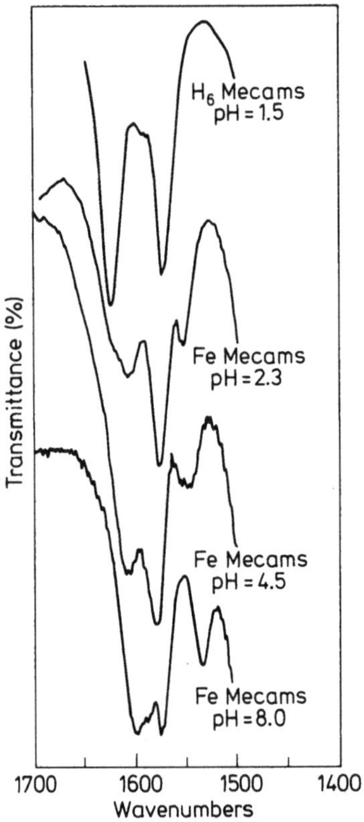

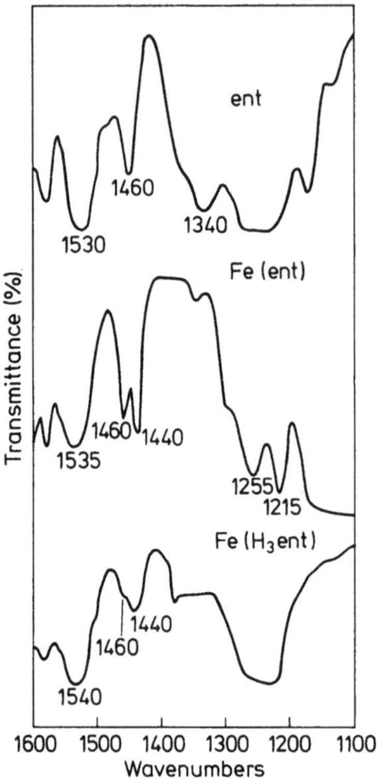

Fig. 12. The FT IR solution spectra of ferric MECAMS as a function of pH in D_2O solution. The infrared band near 1608 cm^{-1} is assigned to metal bonded catechol arms. Only at low pH values does a band grow in near 1628, which appears at the same place as the free ligand and corresponds to the dissociation of one or more catechol arms from the metal complex

Fig. 13. The solid state FT IR spectra of enterobactin (top), red $[Fe(ent)]^{3-}$ (middle) and blue $[Fe(H_3ent)]^0$ (bottom)

analogs) can be explained as follows: at pH 7.4 the predominant species is the dark red $[Fe(ent)]^{3-}$ complex. At pH 4.89 the solution contains a 1:1 mixture of $Fe(ent)]^{3-}$ and the monoprotonated $[Fe(Hent)]^{2-}$, with a small amount of the diprotonated species. By pH 3.15 there is little $[Fe(ent)]^{3-}$ present and the solution contains a 1:1 mixture of $[Fe(Hent)]^{2-}$ and $[Fe(H_2ent)]^-$. The species distribution curves of Fig. 14 illustrates these transformations. As the acidity is increased, the complex is further protonated, which results in precipitation of the dark purple, neutral $[Fe(H_3ent)^0$. Depending upon the total enterobactin concentration, this reaction proceeds between pH 3 and 4. At pH values greater than 4 there is no evidence for an uncoordinated catechol arm. However, at lower pH the band at 1626 cm^{-1} indicates that a catechol arm has dissociated from the metal ion, which remains Fe(III) [130]. There has been some question about the nature of the purple solid; this will be discussed in detail in chapter V.

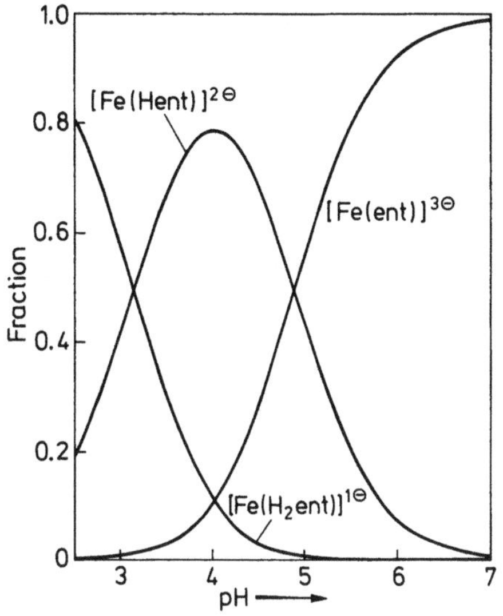

Fig. 14. Species distribution of $[Fe(ent)]^{3-}$, $[Fe(Hent)]^{2-}$ and $[Fe(H_2ent)]^{1-}$. Since the solubility of the purplish blue solid $[Fe(H_3ent)]^0$ is unknown, the extrapolation of the species distribution at low pH to solutions with relatively high total iron concentrations cannot be made accurately

4 Affinity of Siderophore Ligands and Their Analogs for Metal Ions Other than Fe(III)

Metals such as Al(III), Cr(III), Ga(III), and In(III) are direct analogs of ferric ion in that their size and charge (and preferred coordination number for oxygen ligands) are very similar. Thus the other trivalent metal ions follow, at least in part, the same metabolic pathways as iron [131–134]. In particular, the siderophores and synthetic analogs are excellent complexing agents for these metals. Of course differences in their behavior [particularly the kinetic inertness of Cr(III) or the relative instability of a divalent oxidation state for all of these analog ions] can make a critical difference in their actual physiological behavior. In addition, other metal ions such as Pu(IV) and Th(IV) which are larger but resemble ferric ion in their charge/radius ratio [118,135] can also be entrained in certain biological processes involving iron transport in biological systems. Both catechol and, particularly, hydroxamate ligands also form strong complexes with the smaller bivalent metal ions such as Cu(II) [136].

Because of their high relative selectivity for ferric ion, the siderophore ligands have been used either directly (e.g., DFO) or as models in the chelation therapy of iron overload in man. The necessity of such specificity is illustrated by the relative toxicity of chelating agents such as EDTA and DTPA [136], which bind relatively indiscriminately and strongly to biologically important metal ions such as Zn(II), Cu(II) and, in particular, Ca(II). Thus the relative specificity for Fe(III) of catecholate ligands is an important parameter. This has been evaluated by potentiometric titration of solutions containing several biologically important metals and the sulfonated polycatecholate ligands: 4-LICAMS, MECAMS, and 3,4,3-LICAMS [126]. For all the ligands studied the following relative stabilities were observed (see Table VII):

$$\mathrm{Fe(III)} \gg \mathrm{Cu(II)} > \mathrm{Zn(II)} > \mathrm{Ni(II)} \sim \mathrm{Co(II)} > \mathrm{Mg(II)} > \mathrm{Ca(II)}$$

These polycatechoylamide ligands possess great selectivity for ions of high charge to ionic radius ratios such as Fe(III). There is in general a good correlation between the pM values of metal complexes by these catechol ligands and the charge/radius ratios of metal ion (Fig. 15).

The radionuclides gallium-67 and indium-111 have been used extensively as tumor imaging agents [137,138]. One of the limitations of this method is the high background radiation level. This is caused by the distribution of non-tumor deposited ^{67}Ga in the liver, spleen and blood via iron metabolism [134]. Hence it would seem that mobilization of gallium from transferrin would be the most successful mechanism of metal decorporation. Gallium occupies the iron binding sites in transferrin and ferric ion will displace gallium from these sites [139,140]. Transferrin binds iron much more strongly than it binds gallium, by a factor of 400 for K_1^* and a factor of 200 for K_2^* [154].

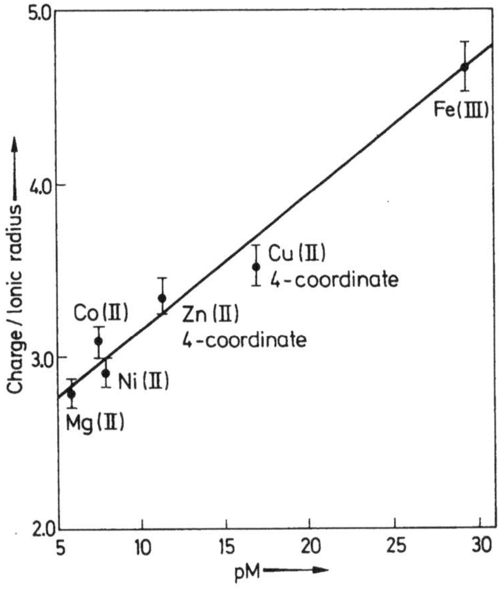

Fig. 15. Graph of charge to ionic radius ratio vs. pM for the enterobactin analog MECAMS

Table VII. Equilibrium free metal ion concentrations expressed as pM[a]

	4-LICAMS[b]	MECAMS[b]	3,4,3-LICAMS[b]	EDTA[c]	DTPA[c]	DFO[d]
Cu(II)	13.6	16.9	14.7	16.9	18.2	11.8
Zn(II)	8.3	11.3	8.7	14.6	15.1	7.2
Ni(II)	6.8	8.0	7.2	16.7	17.0	7.0
Co(II)	6.5	7.7	7.0	14.5	16.0	6.5
Mg(II)	6.0	6.0	6.0	7.0	6.4	6.0
Ca(II)	6.0	6.0	6.0	8.8	7.6	6.0
Fe(III)	23.3	29.3	31.1	22.2	24.7	26.6

[a] $pM = -\log [M(H_2O)_x^{n+}]$; calculated for 10 μM ligand, 1 μM metal, pH 7.4 at 25 °C, and 0.1 M KNO_3. [b] Reference [126]; [c] Reference [216]; [d] References [125,215].

Using the synthetic iron sequestering agents MECAMS and 3,4-LICAMS, the stability constants of the corresponding Ga(III) and In(III) complexes have been determined by spectrophotometric competition experiments between $Fe(MECAMS)^{6-}$ and $Ga(EDTA)^{-}$ (see Tables VII and VIII)[141]. A comparison of the pM values for Ga(MECAMS) and Ga(3,4-LICAMS) shows that these ligands are at least a thousand times more effective than transferrin at binding gallium. In the same paper the metal complex protonation of these compounds has been studied. The data show that the relative stability of the Ga(III) salicylate coordination is less than Fe(III), such that two-proton reactions to give catecholate arm dissociation occurs for even some multidentate catechoylamide ligands.

Table VIII. pM[a] values for selected metal ion sequestering agents

	Fe^{3+} [b]	Ga^{3+}	In^{3+}	Ref.[c]
HBED[d]	31.0	30.9		216)
MECAMS	29.4	26.3	27.4	141)
3,4-LICAMS	28.5	26.0	26.5	141)
EHPG[e]	26.4	23.5		216)
DTPA[f]	24.7	22.8	25.9	216)
Transferrin[g]	23.6			221, 222)
EDTA[h]	22.2	21.6	0.7	220)
TIRON[i]	19.5	19.4		219)
DMBS	19.2	16.1	15.1	141)
Hydroxide[j]	19.4[k]	17.8[k]	17.7[k]	216)

[a] Conditions: $[M^{3+}]_T = 1 \times 10^{-6}$ M; [ligand] $= 1 \times 10^{-5}$ M; pH 7.4; pM $= -\log [M^{3+}(H_2O)_6]$. [b] pM values reported previously in Ref. [123]. [c] pM values for Ga^{3+} and In^{3+} calculated from stability constants in these references. [d] N,N'-bis(2-hydroxybenzyl)ethylenediamine-N,N'-diacetic acid. [e] Ethylene-1,2-bis(2-hydroxyphenyl)glycine. [f] Diethylenetriaminepentaacetic acid. [g] $[HCO_3^-] = 0.024$ M. [h] Ethylenediamintetraacetic acid. [i] 1,2-Dihydroxy-3,5-disulfonatobenzene. [j] pM values are given for $-\log [M^{3+}]_T$, representing all soluble hydrolyzed species. [k] Values of $\log K_{SP}$ used: Fe^{3+}, −38.8; Ga^{3+}, −37; In^{3+}, −36.9.

V Electron Transfer within Siderophores and the Iron Oxidation State

One probable mechanism for the release of iron from siderophores to the agents which are directly involved in cell metabolism is enzymatic reduction to the ferrous state. Due to the very low affinity of hydróxamate and catecholate siderophores for Fe(II), the reduction converts the tightly bound ferric ion to the ferrous complex, which is unstable with respect to protonation and dissociation at neutral pH or below. Therefore comparison of siderophore complex redox potentials with those of physiological reductants can be very useful for the clarification of the mechanism of iron metabolism. Table IV shows the redox potentials [obtained by cyclic voltammetry (see Fig. 18)] of the siderophores tested so far. The values of all of the hydroxamates are within the range of known biological reductants such as *Chromatium* ferredoxin [—490 mV (NHE) or NADH-dehydrogenase [—300 mV (NHE)] [142].

The redox potential determination of tricatecholates was limited to the high pH regimen (~ pH 10). At lower pH proton transfer is involved with the reduction process, leading to irreversible waves in the cyclic voltammograms and preventing direct determination of pH 7 potentials. Based on the observed dependence of the half wave potential with pH it is possible to estimate the pH 7 values. The estimated redox potentials at pH 7 for both enterobactin and parabactin are very high, indicating that intracellular release of iron can only occur by microbial transformation of the irreducible form into a reducible one (see Table IV). However, these potentials are very pH dependent, dropping to +170 mV for ferric enterobactin at pH 4 [130].

Parabactin A, derived from hydrolysis of parabactin (the oxazoline group) exhibits a physiological reducibility. Therefore, oxazoline ring cleavage [148] may be a part of metabolic iron removal in *Paracoccus denitrificans*. A similar process was initially proposed and broadly accepted for the utilization of enterobactin [144]. An esterase was found in the cell extracts of *E. coli* which cleaved the ester backbone of ferric enterobactin [145] and desferri-enterobactin [146]. In addition, it was demonstrated that

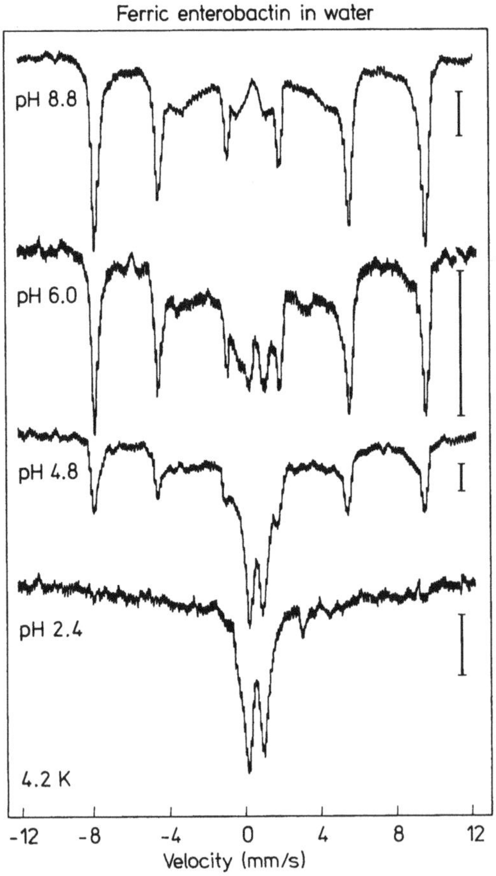

Fig. 16. Mössbauer spectra of ferric enterobactin as a function of pH at 4.2 °K in a magnetic field with the $H^{app} = 60$ mT parallel to the γ ray source. All samples were in frozen solution form. The vertical bars indicate 1 % absorption. With decreasing pH a fast relaxing Fe(III) high-spin species arises at about $\delta = 0.5$ mm/s

the hydrolysis product (dihydroxybenzoylserine, DHBS) is not an intermediate in the biosynthesis [95] of enterobactin. Therefore a mechanism of enterobactin-mediated iron uptake was proposed which involved an initial ester cleavage in order to metabolize the iron [144]. This model fits well the redox potential data, which show a potential for the ferric DHBS complex that falls well within the range of biological reductants.

It was therefore surprising that uptake studies with synthetic analogs which lacked ester units and therefore could not be hydrolysed like enterobactin showed growth response by *B. subtilis* [105] and by *E. coli* [106, 107, 110]. These observations are not consistent with such a simple picture of iron *release* via the esterase. The question of the possible role of an internal redox reaction in the release of iron from ferric enterobactin has been the subject of some controversy in the literature. It is well known that many polyphenol ligands (including catechols) have oxidation potentials for the formation of quinones which are similar to the ferrous/ferric redox couple. For example, the reduction potentials for ortho-quinone/catechol and Fe(III)/Fe(II) are 0.792 and 0.770 V, respectively [146a]. Indeed, the oxidation equilibrium of ferric ion and catechol in acid solution has been carefully studied [147]. However, at higher

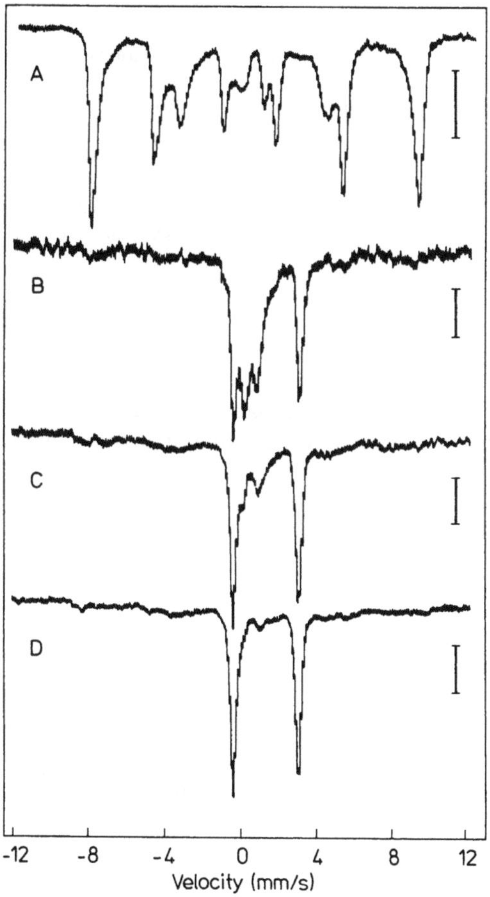

Fig. 17. Mössbauer spectra at 4.2 °K in a magnetic field with the H^{app} = 60 mT parallel to the γ ray source of ferric enterobactin in methanol as a function of pH: (A) pH 9.2, (B) pH 4.0, (C) pH 2.3, (D) pH 1.2. The vertical bars indicate 1% absorption

pH the complexation of ferric ion by catecholate ligands so shifts the redox potential that the internal redox reaction becomes negligible. Several papers nevertheless have appeared which have assumed an intrinsic electron transfer from iron to catecholate ligands low to neutral pH [147–150], including an interpretation of the precipitation of the purple ferric enterobactin protonated complex [151] as such an intrinsic redox reaction. Yet careful Mössbauer spectroscopic study [129] of enterobactin has shown that in aqueous solution there is no chemically significant amount of ferrous ion produced (Fig. 16). However, this reaction does occur at low pH in methanol (see Fig. 17). In the light of the protonation equilibria and redox potential data available

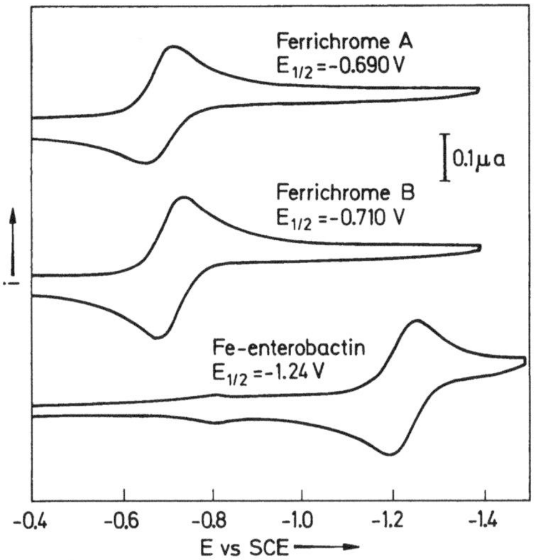

Fig. 18. Cyclic voltammograms of: (top) ferrichrome A (pH 8); (middle) ferrioxamine B (pH 8); (bottom) ferric enterobactin (pH 10.5). All are in 1 M KCl, 0.05 M sodium borate/0.05 M sodium phosphate buffer. All cyclic voltammograms were at hanging mercury drop electrode with 100 mV/sec scan rate

for the ferric tricatecholate complexes, two possible mechanisms of iron removal from enterobactin and microorganisms now remain which do not involve ligand hydrolysis:

1) The internal reduction of ferric ion by the enterobactin catechol groups in a *non*-aqueous environment (perhaps a lipid bilayer of the membrane) which could facilitate the acquisition of the metal ion by ligand exchange.

2) A locally low pH region in the cellular uptake pathway of the ferric enterobactin complex with a resultant shift in the redox potential which would make it available via reduction.

A recent study suggests a reductive removal of iron from enterobactin at pH lower than 6 by glutathione [152]. Mössbauer spectra of frozen *E. coli* cells after ^{57}Fe-enterobactin uptake show an accumulation of the complex and no evidence for the mechanism described as (1) above [153]. However, the occurrence of a fast-relaxing Fe(III) high-spin species is observed, which may be attributed either to an iron storage protein or to a salicylate form of the ferric enterobactin complex, thus indicating a low pH region in the cell (presumably the periplasm).

VI Iron Exchange Kinetics of Siderophores

One crucial mechanism of iron incorporation by organisms is the exchange of the metal from the carrier to another molecule. This may be achieved either by reductive removal or by the involvement of a superior chelating agent. If ligand exchange is part of the iron transport process, the exchange kinetics can be the rate limiting process for uptake. Of course while thermodynamic stability constants determine the equilibrium distribution, they say nothing about the rate of ligand exchange, which is highly important *in vivo*. A similar problem arises in judging the efficacy of sequestering drugs in iron removal from transferrin and ferritin. This aspect is reported elsewhere [89, 102, 115, 155].

Much has been learned recently about the kinetics of aquation and formation in acidic media of desferrioxamine B [128, 156, 157] and model mono(hydroxamato)iron(III) complexes [158–160]. In early papers on chelate exchange of ferric siderophore complexes [16–163] it has been assumed that the ferric ion is readily exchanged, due to the known kinetic lability of high-spin ferric ion binding. However, in order to obtain more detailed information kinetic studies of two types have been performed [164, 165]:

1) The exchange of iron between two ferric siderophores has been monitored using $^{55}Fe^{3+}$ labeling techniques. That is, the kinetics of the equilibrium: $^{55}FeL + FeL' \rightleftharpoons FeL + {}^{55}FeL'$.

2) The kinetics and mechanism of iron removal from siderophore complex to a free ligand has been examined using spectrophotometric or $^{55}Fe^{3+}$ labeling techniques.

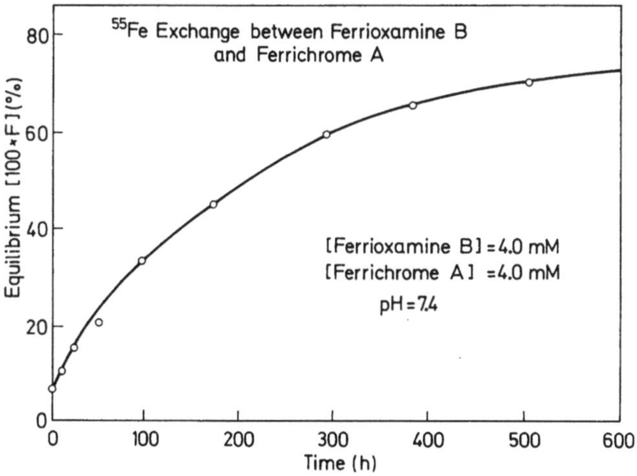

Fig. 19. Reaction profile for exchange of iron between ^{55}Fe-labeled ferrioxamine B and ferrichrome A. The quantity F is the functional degree to which equilibrium has been reached

Figure 19 shows the time-dependent ^{55}Fe exchange between two ferric hydroxamate siderophore complexes: ferrioxamine B and ferrichrome A. [165] This process is an extremely slow pseudo-first order reaction under the conditions described, which is only 50% completed after *220 hours* of incubation at 25 °C and pH 7.4. Lowering

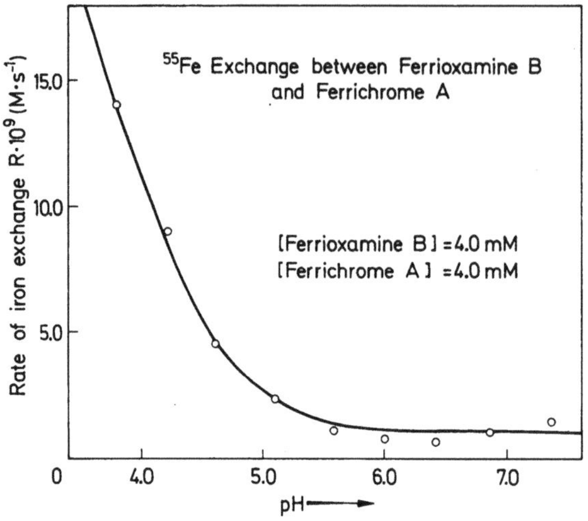

Fig. 20. The dependence of the ^{55}Fe exchange rate, R, on pH

the pH greatly accelerates the iron exchange rate as a function of increasing hydrogen concentration (see Fig. 20). The ^{55}Fe studies of hydroxamate complex-free ligand exchange (see Table IX) show that an approximate linear-free-energy relationship exists between the equilibrium competition constant and the second-order rate constant for iron removal from ferrichrome, ferricrocin and coprogen to ferrioxamine B [164].

In contrast to the slow exchange observed for tris hydroxamato complexes, iron exchange between ^{55}Fe labeled 3,4-LICAMS and ferrioxamine B displays relatively

Table IX. Second-order rate constants for hydroxamate/hydroxamate ligand exchange in siderophore complexes

$$^{55}\mathrm{FeL}^{3-n} + \mathrm{H_4DFO^+} \underset{k_{-2}}{\overset{k_2}{\rightleftharpoons}} {}^{55}\mathrm{FeHDFO^+} + \mathrm{H_3L}^{3-n}$$

FeL	k_2 $(M^{-1}\cdot s^{-1})\times 10^2$	k_{-2}[a] $(M^{-1}\cdot s^{-1})\times 10^2$	$\log \beta_{110}$[b]	$\log K_{comp}$[c]	$\log (k_2/k_{-2})$
Ferrichrome	6.3(2)	0.67(3)	29.1[d]	1.1[d]	1.0
Ferricrocin	1.2(4)	0.41(9)	30.4[e]	—0.1[e]	0.5
Coprogen	0.46(2)	4.2(3)	30.2[e]	—1.4[e]	—1.0
Ferrichrome A	0.089(6)	2.7(9)	—	—	—1.5
Ferrioxamine B	—	—	30.1[f]	0	—

[a] pH 7.4 (0.1 M Tris HCl); $\mu = 0.2$ M (KNO_3); T = 25.0 °C. [b] $\beta_{110} = \dfrac{[\mathrm{FeL}^{3-n}]}{[\mathrm{Fe}^{3+}][\mathrm{L}^{n-}]}$

[c] $K_{comp} = \dfrac{\beta_{110}^{FeHDFO}\beta_{013}^{L}}{\beta_{110}^{FeL}\beta_{013}^{DFO}}$; $\beta_{013}^{L} = \dfrac{[\mathrm{H_3L}^{3-n}]}{[\mathrm{H}^+]^3[\mathrm{L}^{n-}]}$.

[d] $\mu = 0.1$ M; T = 20.0 °C; Refs. [125, 157].

[e] $\mu = 0.1$ M; T = 25.0 °C; Ref. [214]. [f] $\mu = 0.1$ M; T = 25.0 °C; the value of β_{110} used in the analysis was measured at 25.0 °C and $\mu = 0.1$ M, Pecoraro, V. L., Raymond, K. N., unpublished results.

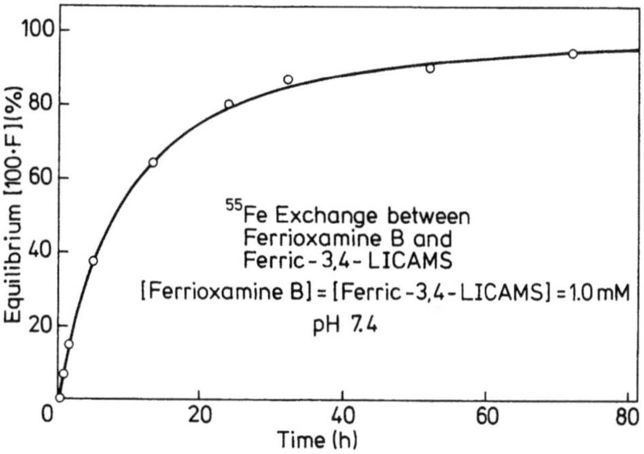

Fig. 21. Kinetic profile for $^{55}Fe^{3+}$ scrambling between ferric (3,4-LICAMS) and ferrioxamine B

rapid kinetics (see Fig. 21) [166]. A half-life of approximately 10 hours was observed for equimolar concentrations (1.0 mM) of the competing complexes with no excess free ligand.

From a mechanistic point of view, the characteristic feature of siderophore iron exchange processes is the unraveling of one ligand from the inner coordination sphere, concomitant with complexation by the incoming ligand. This occurs via the formation of ternary complex between metal ion and entering and leaving ligands. The general features of these exchange processes conform to ideas formulated in earlier studies of multidentate ligand exchange [167].

Table X. Kinetic data for the reaction of hydroxamate complexes with 3,4-LICAMS. Data are taken from Ref. [164]

FeL	FeL + 3,4-LICAMS ⇌ Fe(3,4-LICAMS) + L		
	$k_1^{obs\,a}$ $(s^{-1}) \times 10^3$	$\log K_{comp}$ [b]	β_{110} [c]
Tris(acetohydroxamato)iron(III)	>700	6.5[d]	27.8[e]
Ferric rhodotorulate	55	2.1[f]	31.2[g]
Ferric dimerate	19	—	—
Coprogen	4.1	1.8	30.2
Ferrioxamine B	2.0	3.2	30.1
Ferrichrome	0.53	4.3	29.1[h]
Ferricrocin	0.11	3.1	30.4
Ferrichrome A	0.016	—	—

[a] $[FeL]_T = 0.10$ mM; $[3,4\text{-LICAMS}]_T = 1.0$ mM; pH 7.4; T = 25.0 °C; μ = 0.20 M.

[b] $K_{comp} = \dfrac{\beta_{110}^{FeLICAMS}\beta_{013}^{L}}{\beta_{110}^{FeL}\beta_{013}^{LICAMS}}$; T = 25.0 °C; μ = 0.10 M. [c] $\beta_{110} = \dfrac{[FeL^{3-n}]}{[^-Fe^{3+}][L^{n-}]}$.

[d] $K_{comp} = \dfrac{\beta_{110}^{Fe\text{-}LICAMS}(\beta_{011}^{L})^3}{\beta_{130}^{FeL_3}\beta_{013}^{LICAMS}}$. [e] β_{130}. [f] $K_{comp} = \dfrac{\beta_{110}^{Fe\text{-}LICAMS}(\beta_{012}^{L})^{3/2}}{(\beta_{230}^{Fe_2L_3})^{1/2}\,\beta_{013}^{LICAMS}}$

[g] $(\beta_{230})^{1/2}$. [h] T = 20 °C.

Taking 3,4-LICAMS as representative for catechol ligands it was demonstrated that the free ligand catechols can more effectively compete — kinetically as well as thermodynamically — for hydroxamate bound iron at lower concentrations and at higher pH in comparison to hydroxamates (see Table X)[164,166]. The kinetics reveal first order dependence on the concentration of both hydroxamate and tri-catechoylamide. The pH dependence is more complicated as the data in Fig. 22 (for a MECAMS study) indicate. Acid catalysis is evidenced at low pH — again reflecting the general role that hydrogen ion plays in these iron exchange and removal studies involving siderophores.

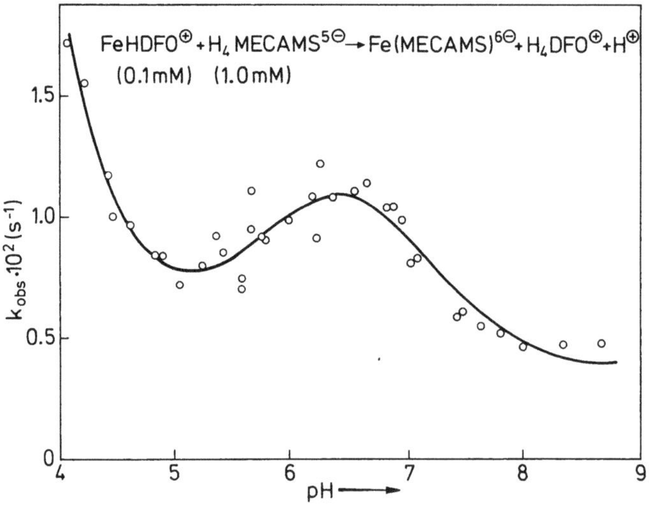

Fig. 22. The dependence of the observed first-order rate constant for iron removal from ferrioxamine B

In summary, the kinetics and mechanism of iron exchange between siderophores have revealed that iron bound in these complexes is *not* in general readily exchanged. However, catecholate complexes display more rapid reaction kinetics compared to their hydroxamate counterparts.

VII Stereochemistry of Siderophores

Siderophores may in general form both geometrical and optical isomers in solution. However, the high-spin d^5 electronic configuration of the iron as shown by EPR[168,169], Mössbauer[170-172] and magnetic susceptibility measurements[173], rules out any crystal field stabilization energy and makes the complexes relatively labile with respect to isomerization and ligand exchange in aqueous solution. In addition there are no spin-allowed d-d optical transitions, thus the CD and UV-VIS spectra, which are dominated by charge transfer bands are not *a priori* interpretable on the basis of theoretical considerations, although empirical similarities have evolved (see Table XI).

Table XI. Spectroscopic properties of the siderophores and analogs. All solutions are aqueous except for the benzhydroxamate complexes (acetone)

		CD bands, nm (Δε)		Refs.
		Δ	Λ	
Ferrichrome	425 (2895)		360 (−3.7)	181, 223
			465 (2.4)	
Ferrichrome A	440 (3360)		330 (−3.9)	192, 224
			365 (−2.7)	
			465 (3.2)	
Ferrichrysin	430 (3020)			225
Ferricrocin	434 (2460)		290 (−3.78)	214
			360 (−1.62)	
			450 (2.47)	
Ferrioxamine B	428 (2800)			26
Ferrioxamine E	430 (2750)			26
Coprogen	434 (2820)	375 (2.1)		214
		474 (−1.26)		
N,N′,N″-triacetylfusarinine		370 (3.25)		180
		467 (−2.04)		
Mycobactin P	445 (3780)	Δ-*cis*(crystals)		26, 32
Fe_2RA_3	425 (2700)	372 (2.73)		36, 188
		464 (−1.41)		
$Fe(benz)_3$	435 (4910)	350 (2.3)	350 (−2.8)	192
		452 (−1.5)	455 (1.1)	
Aerobactin	398 (2170)		415 (−0.12)	30
pH 7 $[Fe(H,L)]^{3-}$			574 (0.25)	
			668 (−0.10)	
Enterobactin	495 (5600)	Δ-*cis*		46, 178
				226, 227
Agrobactin	505 (4100)			33
Parabactin	512 (3300)			33
Pseudobactin	400 (15000)		400 (2.0)	34
			436 (−0.8)	
			502 (0.3)	
Pseudobactin A	400 (2000)			35

These "disadvantages" of iron for any stereochemical investigation of siderophore complexes have been circumvented by substituting the ferric ion with kinetically more inert d^3 chromium(III) or d^6 rhodium(III) ions, which have the same charge and size as iron(III) — yet have a d-electron configuration granting significant crystal field stabilization energy for kinetic inertness and have well characterized d-d transitions with distinct UV-VIS and CD spectra.

Raymond and co-workers have synthesized and separated optical and geometrical isomers of simple tris hydroxamate chromium(III) and tris phenolate chromium(III) or rhodium(III) complexes and assigned the absolute configurations of these isomers based on criteria such as: chromatographic behavior due to differences in dipole moment, theoretical symmetry considerations, and X-ray crystallographic data [174]. The absolute configuration of isomers of chromium(III) complexes of ferrichrome, ferrichrysin [175], ferrioxamine B and D_1 [176], rhodotorulic acid [177], enterobactin [178], aerobactin [30] and thioformin [179] could then be assigned by comparison of their visible and CD spectra with those of the models. If single crystals of the ferric complexes can be obtained, a direct analysis of the CD bands is possible. Comparing the crystal structure with the CD spectrum of the crystalline solid and the dissolved compound, the bands can be assigned. With this method, van der Helm et al. confirmed the configuration of N,N′,N″-triacetylfusarinine and ferrichrome in solution as Δ and Λ, respectively [180, 181].

The isomers of kinetically inert chromium siderophore complexes have been used as biological probes to elucidate microbial siderophore uptake systems, answering the following questions:

1) What are the limits of specific recognition in receptor dependent uptake; is there any discrimination between different isomers?

2) What are the mechanisms of uptake; is the whole complex transported into the cell, or is there removal of iron at the cell membrane by reduction or decomplexation?

1 Use of Inert Complexes as Chemical Probes

Since most hydroxamate siderophores are hexadentate ligands with three unsymmetrical bidentate functional units, 16 geometrical and optical isomers are theoretically possible for the metal complexes, Raymond and co-workers have proposed a nomenclature [182]. The isomers are names as follows (see Fig. 23): looking down the pseudo C_3 axis, Δ isomers have a right-handed propeller configuration about the metal ion, while the Λ isomers have the left-handed propeller configuration. The molecule is oriented such that the sequence of hydroxamate (designated 1, 2, and 3 for each ligand) corresponds to the rotation direction. If the ring 1 has the carbon atom of the hydroxamate group below the nitrogen, it is denoted ″C:. If the reverse is true, it is called "N". For rings 2 and 3 each is called *cis* or *trans* depending upon whether it has the same or opposite relative orientation with respect to the coordination axes as does ring 1. Figure 23 shows the one set of geometrical isomers of coprogen which are not redundant, as derived from a molecular model.

Due to the rather complicated stereochemistry of siderophores, simple hydroxamate and phenolate chromium(III) complexes, for which only four isomers are possible, have been used as models for the correlation between absolute configuration and UV-VIS and CD spectra. Tris(benzhydroxamate)chromium(III) has been separated into the *cis* and *trans* geometrical isomer as confirmed by X-ray structural analysis [183].

Three fractions of tris(N-methyl-1-menthoxyacetohydroxamato)chromium(III) were separated by chromatographic techniques and assigned to be Δ-*cis* and Λ-*cis*, while the partially Δ- and Λ-*trans* mixture was obtained as one fraction. The assignment was based on the chromatographic properties of the isomers and a comparison of their visible and CD spectra with literature data which included octahedral complexes of various transition metals with asymmetric ligands [184].

Fig. 23. Eight geometrical isomers of a ferric tri-chelate complex, involving three unsymmetrical bidentate ligands attached to an asymmetric backbone (e.g. coprogen). The optical conformation at the metal is Δ. Not shown is the set of eight Λ diastereomers. See text for nomenclature ▶

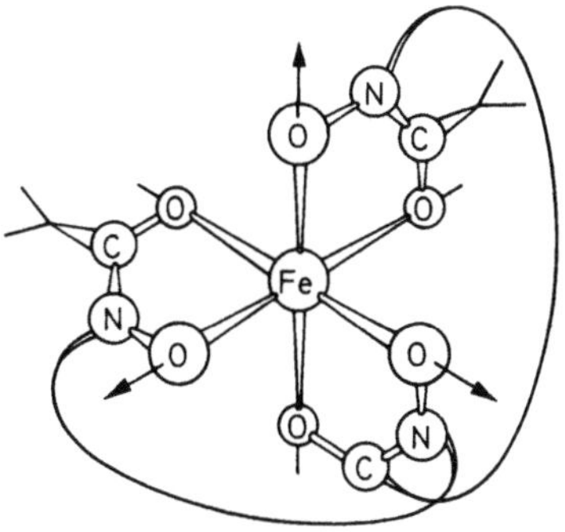

Δ-C-*cis*, *cis*

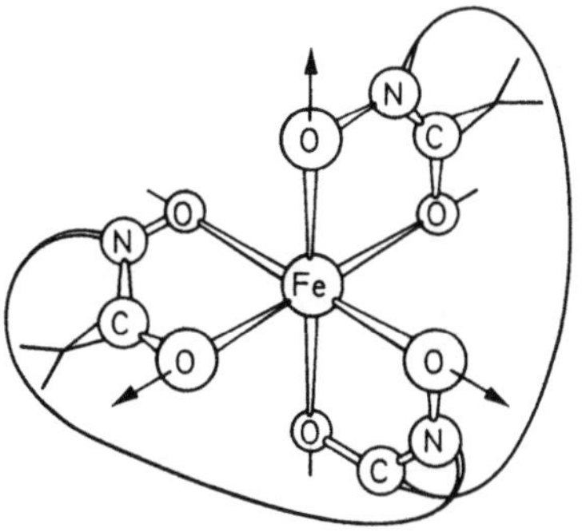

Δ-C-*cis*, *trans*

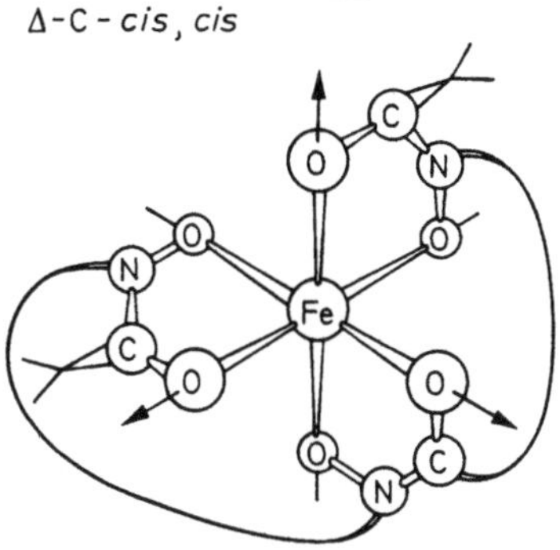

Δ-N-*cis*, *cis*

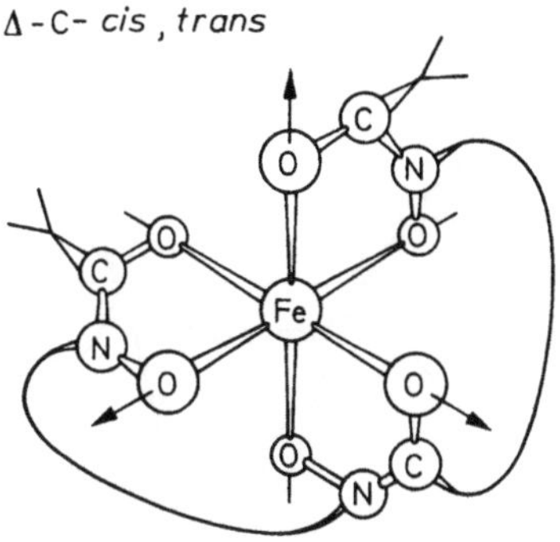

Δ-N-*cis*, *trans*

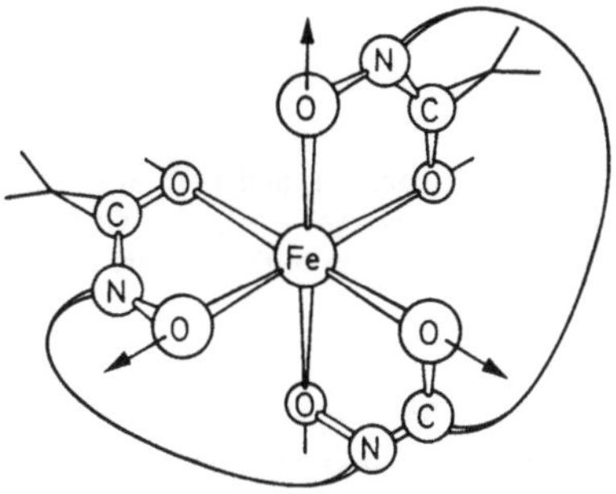

Δ-C-*trans*, *cis*

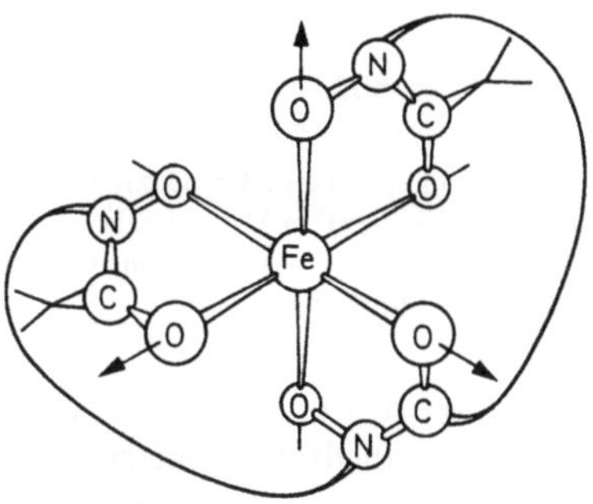

Δ-C-*trans*, *trans*

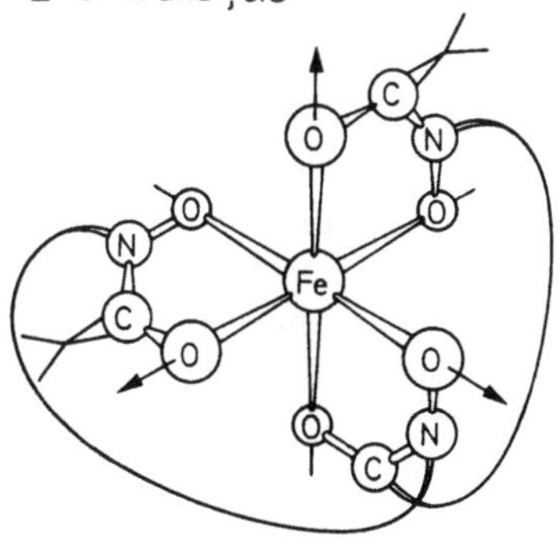

Δ-N-*trans*, *cis*

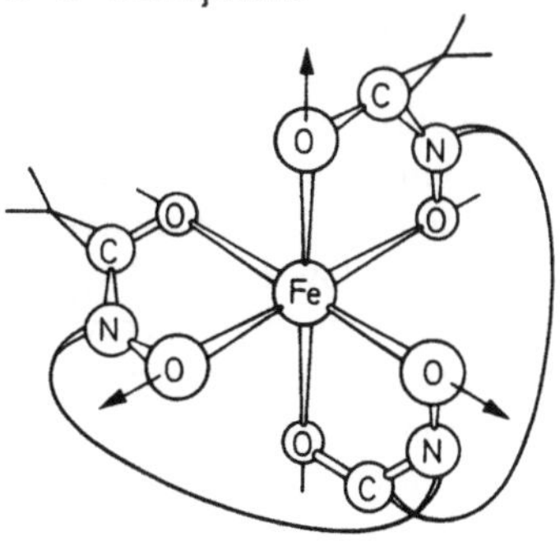

Δ-N-*trans*, *trans*

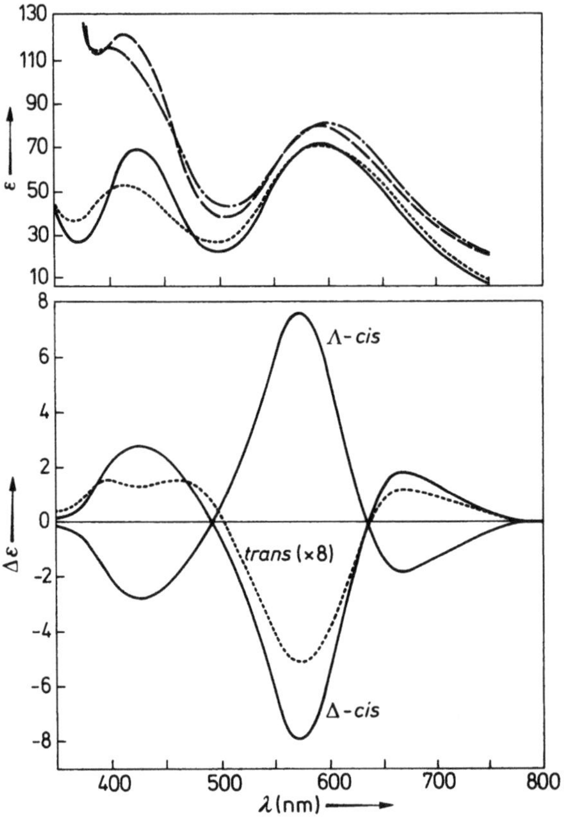

Fig. 24. Absorption spectra of $Cr(benz)_3$ in 17% $CH_3OH—CHCl_3$ solution, and both absorption and CD spectra of $Cr(men)^3$ in 3% $CH_3OH—CHCl_3$ solution: *cis*-$Cr(benz)_3$ (— — —), *trans*-$Cr(benz)_3$ (— · —), *cis*-$Cr(men)_3$ (———), and *trans*-$Cr(men)_3$ (· · ·). The CD spectrum of the mixture of *trans* isomers has been multiplied by eight

The visible spectra of the Cr(III) geometrical isomers have two spin-allowed d-d transitions for octahedral symmetry $^4A_{2g} \rightarrow {}^4T_{1g}$ at higher energy and $^4A_{2g} \rightarrow {}^4T_{2g}$ at lower energy. The absorption maximum of the *trans* isomers is considerably lower at the high energy transition manifold, and in addition the peak maxima are shifted to higher and lower energy as compared to the *cis* isomers.

The interpretation of the CD spectra is based on the following considerations: assuming D_3 point symmetry for simple *cis* tris hydroxamate complexes and pseudo D_3 symmetry for *cis* trihydroxamates, both spin-allowed d-d transitions factor into A and E state. The low energy $^4A_{2g} \rightarrow {}^4T_{2g}$ transition splits into $^4A_2 \rightarrow {}^4A_1(A_2)$ and $^4A_2 \rightarrow {}^4E(E_a)$. The high energy $^4A_{2g} \rightarrow T_{1g}$ transition splits into $^4A_2 \rightarrow {}^4E(E_b)$, and $^4A_2 \rightarrow {}^4A_2$ (which is symmetry forbidden). So the CD spectrum for *cis* tris hydroxamate Cr(III) complexes consists of three bands, due to the E_a, A_2, and E_b transitions. For the interpretation of the CD spectrum the low energy region is especially important. The Λ configuration in the model compounds was found to have a postive CD band for the dominant low energy transition region at 500–600 nm (E_a transition), the Δ isomer had a negative CD band in this region [184]. For the *trans* isomer, which has no symmetry (C_1) the E_b transition at high energy is split, providing an additional criterion for differentiating between *cis* and *trans* isomers [184] (Fig. 24).

2 Absolute Configuration and Separation of Isomers

The molecular structures of ferrichrome [181], ferrichrome A [185], ferrichrysin [186] and alumichrome A [185] have been determined by X-ray crystallographic analysis, which shows in all cases a Λ-*cis* configuration. However an examination of the molecular models of the ferrichromes indicates that both Λ-*cis* and Δ-*cis* isomers could be possible, but not the *trans* isomer, due to the rigidity of the cyclic molecule.

The formation of Λ-*cis* crystals did not necessarily exclude the existence of the other diastereomer in an equilibrium. Attempts to resolve isomers of Cr-desferri-ferrichrome and Cr-desferri-ferrichrysin only resulted in one fraction, which has exactly the same CD spectrum as the model complex Λ-*cis* $Cr(men)_3$, with similar $\Delta\varepsilon$ values of the E_a transition bands (Fig. 27). This indicated, together with crystallographic data, that ferrichrome in solution has an overall Λ-*cis* configuration [145].

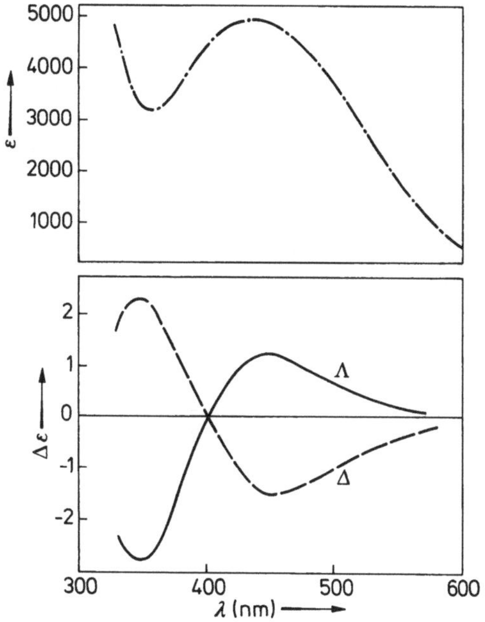

Fig. 25. Visible absorption (— · — · —) and circular dichroism spectra of Λ (———) and Δ (– – –) tris(benzohydroxamato(iron(III) in acetone solution

Ferrioxamines are the only known siderophores with no chiral center, so ferrioxamines are the only siderophores with no intrinsic optical activity of the metal complex. Ferrioxamine E, a cyclic ferrioxamine, crystallizes as a racemic mixture of Δ-*cis* and Λ-*cis* isomers as determined by X-ray crystallography [187]. From an examination of molecular models of ferrioxamine B five enantiomeric pairs of non-redundant isomers are distinguishable. By ion exchange chromatography the more polar *cis*

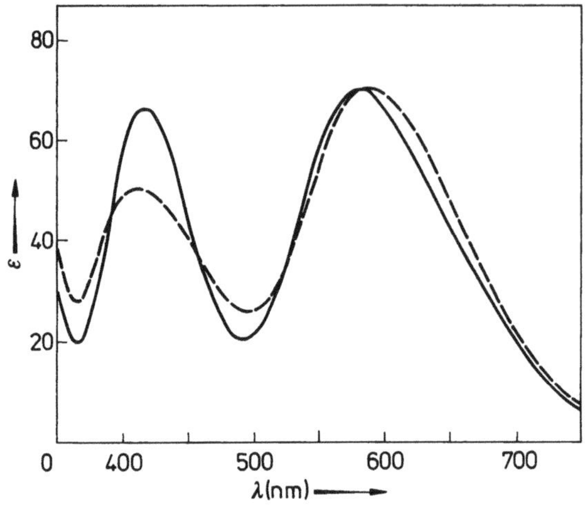

Fig. 26. Absorption spectra of the *cis* isomer and *trans* isomers of chromic desferrioxamine B in aqueous solution: *cis* (———), *trans* (— — —)

isomer could be separated from one or more *trans* isomers [182]. The UV-VIS spectra are very similar to the model $Cr(men)_3$ isomers. The fraction with the lower absorption at higher energy was assigned as one or more *trans* isomers, the other one as *cis* isomer (Fig. 26). Another interesting case is the metal complex of N,N′,N″-triacetylfusarinine in which two optical isomers, Δ and Λ could be crystallized, depending upon the conditions [180].

For a mixed hydroxamate siderophore, mycobactin P, X-ray crystallographic data show that the Fe(III) complex crystallizes in the Δ-*cis* configuration. The absolute configuration of pseudobactin and pseudobactin A is Λ as could be confirmed by crystallographic data as well as by the CD spectra of the iron and the chromium complexes [34, 35].

Rhodotorulic acid (RA), a dihydroxamate siderophore, forms dimeric complexes with iron, aluminium and chromium of the stoichiometry $M_2(RA)_3$ at neutral pH [36, 188]. The coordination chemistry of this siderophore is probably the most complicated of the siderophores. The combination of *cis-trans*, Δ and Λ configurations of two iron centers, connected by three RA molecules, makes 42 non-redundant isomers theoretically possible; each can be simulated by molecular models. Recently three different isomers or mixtures of isomers of Cr_2RA_3 were separated by reversed phase HPLC-chromatography [177]. The visible spectrum of the most abundant fraction corresponds to the *cis* isomer; the two other fractions are very similar to the visible spectrum of the *trans* $Cr(men)_3$ isomer. The CD spectra, in comparison with the $Cr(men)_3$ model complex, show two different optical isomers, assigned as Δ-*trans* and Λ-*trans*. The Λ isomer preparation seems also to contain a certain amount of the Δ configuration. This is the first time that two different, kinetically stable optical isomers have been isolated from the metal complexes of a siderophore [177].

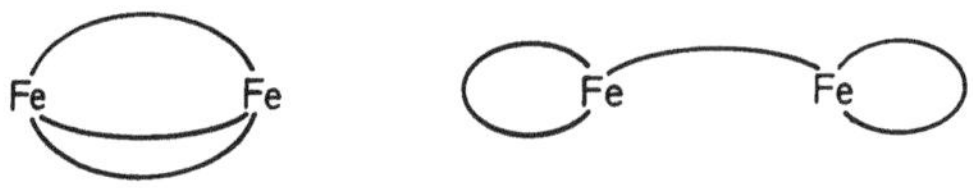

A *B*

Since no crystal structure of $Fe_2(RA)_3$ has been reported so far, it cannot be unambiguously said that the binucleic complex has structure (A) rather than (B).

In an effort to lend support to the triply bridged dimeric structure proposed for ferric rhodotorulate, a series of model ligands, i-$C_3N_7N(OH)C(=O)(-CH_2-)_n$-$C(=O)N(OH)$-i-$C_3H_7$ (n = 3–6, 8, 10) were prepared [86], and their coordination chemistry with Fe^{3+} was examined. In frozen methanol, relaxation effects in the EPR and Mössbauer spectra suggest a dipolar interaction between the ferric ions for the shorter chains (n = 3, 4, 6), thus supporting the proposed triply bridged structure (A). The magnitude of this interaction increases with decreasing chain length; the ferric ions are noninteracting in the complexes with RA and the synthetic ligands with n = 8, 10 [189].

In contrast, the n = 5 complex shows a simple quadrupole doublet in the Mössbauer spectrum (MeOH, 4.2 K) as opposed to the magnetically split spectra observed for the other ferric dihydroxamate complexes [189]. This result implies strong coupling between the iron centers, possibly through an oxygen bridge. The anomalous n = 5 complex crystallizes from methanolic solution and exhibits the stoichiometry: $Fe_2L_2(OCH_3)_2$. The crystal structure obtained [190] shows that both bidentate ligands bind at each iron center. The dimeric *trans* complex is completed by two bridging methoxide groups. The structure explains the spectral anomalies of the n = 5 RA-analog. Moreover, the bridging structure provides additional support for the formulation of Fe_2RA_3 as tribridged species (A), rather than monobridged species (B).

Very recently a hetero-analog of RA, N^1,N^3-(1-hydroxy-2(H)-pyridonecarboxoyl)-diaminopropane has been synthesized (see Fig. 27). The crystal structure of the ferric

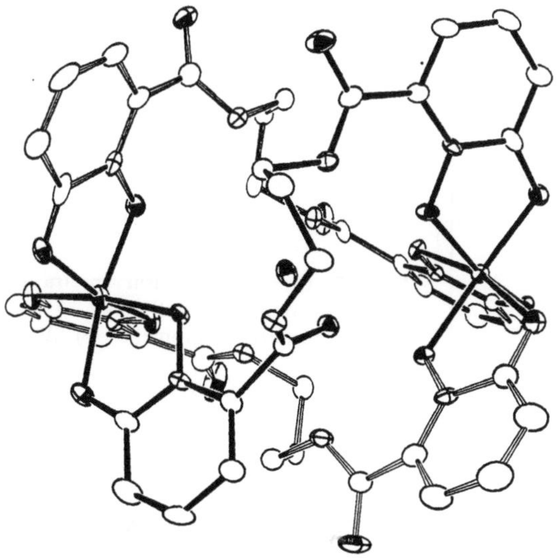

Fig. 27. A perspective view of the 3:2 complex of N^1,N^3-di(1-oxy-2(1H)-pyridinonate-6-carboxyl)-1,3-propanediamine dianion with Fe^{3+}. The thermal ellipsoids are drawn at 20% probability levels — in outline only for carbon atoms, with principle ellipses shown for nitrogen atoms, and with shading for iron and oxygen atoms. Hydrogen atoms have been omitted for clarity. A water molecule occupies the center of the dimer

Fe_2L_3 complex reveals a tribridged structure (A) and a *cis* isomer with respect to the iron octahedra [191]. With these results at hand structure (A) is very likely for the ferric complex of rhodotorulic acid.

As with hydroxamate siderophores, simple tris(catecholato) metallate(III) complexes have served as models for enterobactin. Unlike hydroxamate, catecholate is a symmetric, bidentate ligand. Consequently, there are no geometrical isomers of simple tris(catecholato) metal complexes, and only Λ and Δ optical isomers are possible. However all siderophore catecholates are substituted asymmetrically on the catechol ring, such that geometric isomers may in principle exist. However, in the case of enterobactin molecular models show only the more symmetric *cis* chelate is possible, as the Δ or Λ form.

The simple model complex, tris(catecholato)chromate(III) has been prepared, and complete resolution of the optical isomers was achieved at pH 13. The known crystal structure of $[Cr(cat)_3]^{3-}$ and arguments similar to those for the hydroxamate chromium complexes lead to the assignments of absolute configuration of the CD spectra. It was found that the CD spectra of Δ-$[Cr(cat)_3]^{3-}$ and $[Cr(ent)_3]^{3-}$ are essentially identical, and the mirror image of chromic desferriferrichrome (Fig. 28), which shows that enterobactin has a predominant Δ-*cis* absolute configuration [147]. Unfortunately the usual oxidation sensitivity of the catechol dianion is substantially increased in the chromium complexes, which precludes their use as biological probes [147].

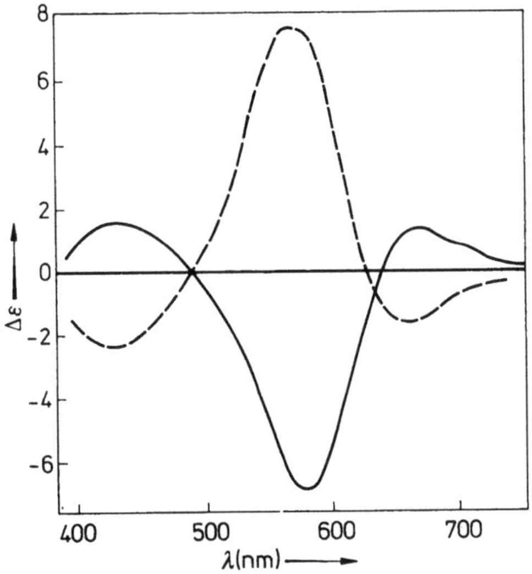

Fig. 28. Circular dichroism spectra of —$(NH_4)_3$[Cr(enterobactin)] (———) and chromic desferriferrichrome (· · · ·) (Δε in $l \cdot mol^{-1}\ cm^{-1}$)

For the ferric siderophore complexes, comparison of the CD spectra of the chromium complexes of ferrichrome and enterobactin with the CD spectra of their iron complexes [and the separation of optical isomers of even ferric(benzhydroxamate)$_3$ complexes in nonaqueous solution [192]] have shown that the same rule applied to the CD spectra for chromium complexes can be used for iron siderophore complexes as well: iron(III) complexes will have a predominant Λ configuration in solution if the CD band in

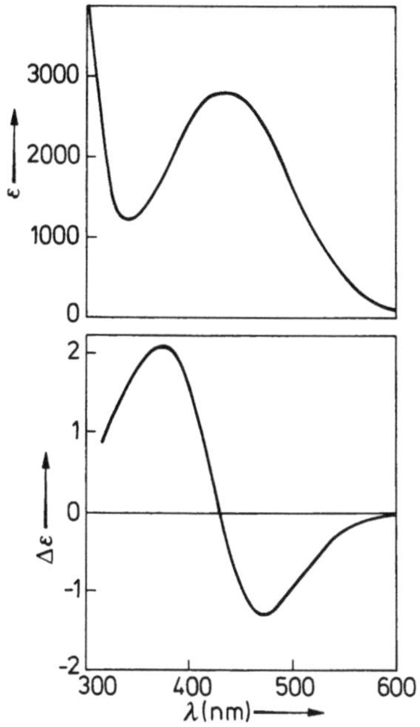

Fig. 29. Electronic and circular dichroism spectra of coprogen. The negative band with maximum at 474 nm indicates a Δ configuration of the metal chelate octahedron

the region of the absorption maximum at 400–500 nm has a positive sign. This despite the fact that the Fe(III) CD spectra are due only to charge transfer bands (Fig. 29). In Table XI the CD spectra of iron complexes in aqueous solution show that all siderophores (except the ferrioxamines, which have no chiral center associated with the ligand) preferentially form one isomer sufficiently that they can be unambiguously classified as Λ or Δ. It is an interesting problem whether the chirality at the metal center for a given siderophore complex can be predicted. It may well be that modern stereochemical computer programs can now make such predictive analysis.

VIII Specificity and Stereospecificity of Microbial Siderophore Uptake

The investigation of microbial iron metabolism embraces the various aspects including: membrane receptor studies [65, 193]; mechanism of uptake [194–196]; intracellular iron removal [196–201] and iron storage [202]; and the genetic regulation of these processes [203]. The coverage of this whole biochemical spectrum would be beyond the scope of this article. Therefore we will focus on the specificity of the siderophore uptake in microorganisms, i.e., the correlation between structure and recognition.

It is generally assumed that siderophore uptake in microorganisms is both a receptor-dependent and energy-dependent process. The enterobactin outer membrane receptor and the ferrichrome receptor of *E. coli* have been isolated by genetic techniques [66, 204, 205]. The isolation of a coprogen binding protein from the fungus *Neurospora crassa* [206], and the investigation of specific and non-specific binding sites of the cytoplasmic membrane also suggest a receptor-mediated ferric siderophore uptake process in this strain [207].

Specific recognition and transport of a siderophore may be dependent on different parts and structural features of the molecule:

1) the chirality or geometric coordination about the metal center;
2) the geometry and chirality of the backbone;
3) certain peripheral groups;
4) the molecule as an entity.

Recent studies with synthetic siderophores have revealed important correlations between the coordination chemistry and recognition. Growth promotion tests with siderophore auxotroph mutants of *E. coli* demonstrated a clear discrimination between the naturally-occurring Δ-*cis* enterobactin and the synthetic Λ-*cis* enantio-enterobactin. The synthetic analogue, derived from D-serine, did not support growth [208] Moreover, in contrast to enterobactin, the enantiomer showed significantly decreased protection against colicin B, a killer protein which enters the cells via the enterobactin receptor [208,209]. Positive growth promotion was achieved by the use of two different achiral enterobactin analogues, in which the cyclic triester moiety of enterobactin was replaced with a carbocyclic and benzene backbone, respectively [104,110]. These results suggested the primary importance of the catecholate iron protein of the molecule in recognition at the membrane. This hypothesis has been confirmed and detailed by uptake measurements with systematically modified synthetic enterobactin analogues and various *E. coli* strains [106]. Two important results have been obtained. First *E. coli* cells accumulate ferric MECAM at rates comparable to enterobactin. The single structural difference between MECAM and enterobactin is the replacement of the seryl-ester platform by mesitylene. Secondly, the employment of the analogues ((Me_3)MECAM, TRIMCAM and MECAMS, each of them reflecting selective changes of the initial MECAM structure, lead to a drastic reduction of the uptake,

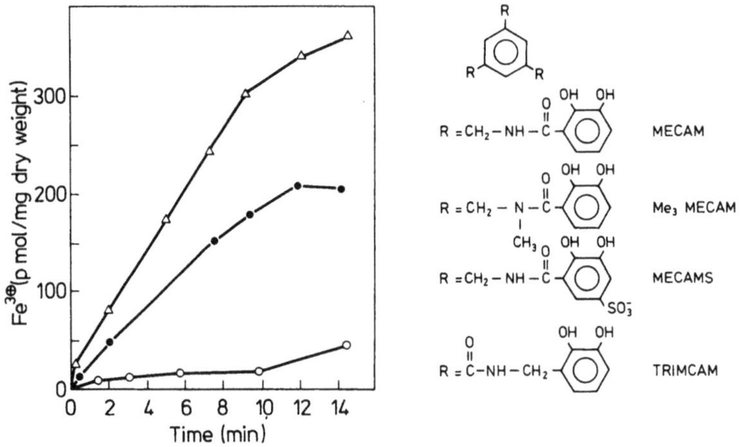

Fig. 30. Uptake of iron(III) into *E. coli* AB2847 as enterobactin (Δ), MECAM (●) and TRIMCAM (○) complexes. The Me_3MECAM is taken up at the same rate as TRIMCAM. The MECAMS show growth promotion with only one out of ten strains tested, Me_3MECAMS is completely growth inactive

which was as low as 10% of the MECAM value. The sulfonated MECAMS was not taken up. The deviant structural features must be responsible for the dropped uptake rate of these compounds (see Fig. 30). In the case of Me_3MECAM the amide nitrogene is alkylated to probe the importance of the amide bond for the recognition while TRIMCAM is a structural isomer of MECAM in which the methylene and the carbonyl functions have been interchanged, thus probing the importance of the α-carbonyl group. With MECAMS, sulfonated at the 5 position of the catecholate ring, steric constraints and the alteration of the electronic character of the catecholate rings are tested. From these results the following inference can be made: the enterobactin uptake receptor recognizes the tricatecholate iron complex propeller and changes in the connecting amide group. However, the cyclic serine-triester backbone is of minor or no importance [106].

Rhodotorula pilimanae, a yeast which produces vast amounts of rhodotorulic acid (RA, Fig. 1), has also been investigated for the specificity of Fe_2RA_3 uptake. Again the goal was to clarify which part of the molecule is important for the recognition: the diketopiperazine rings, the iron centers, or the whole molecule? Uptake measurements have been performed with a series of dihydroxamic acid analogs of RA, in which the hydroxamate groups are separated by carbon chains of 3 to 8, and dimerum acid, a naturally occurring dihydroxamate siderophore. These compounds all deliver iron to *Rhodotorula pilimanae*, however none of the monomeric siderophores such as ferrioxamine B, ferricrocin and coprogen are taken up (Fig. 31). These results indicate that the diketopiperazine ring is not necessary for recognition. Further experiments

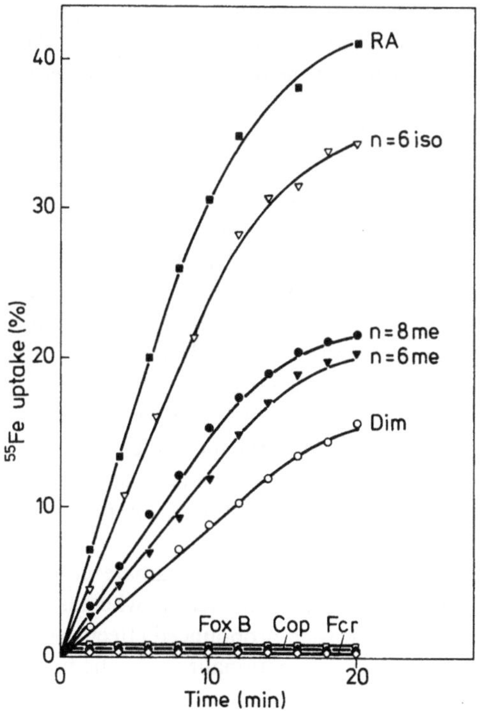

Fig. 31. Time dependent iron(III) uptake mediated by siderophores and synthetic analogs of rhodotorulic acid (RA); n = 6 iso: N,N′-dihydroxy-N,N′-diisopropyloctanediamide; n = 8 me: N,N′-dihydroxy-N,N′-methyldecanediamide; n = 6 me: N,N′-dihydroxy-N,N′-methyloctanediamide; Dim: dimerum acid; Fox B: ferrioxamine B; Cop: coprogen; Fcr: ferricrocin

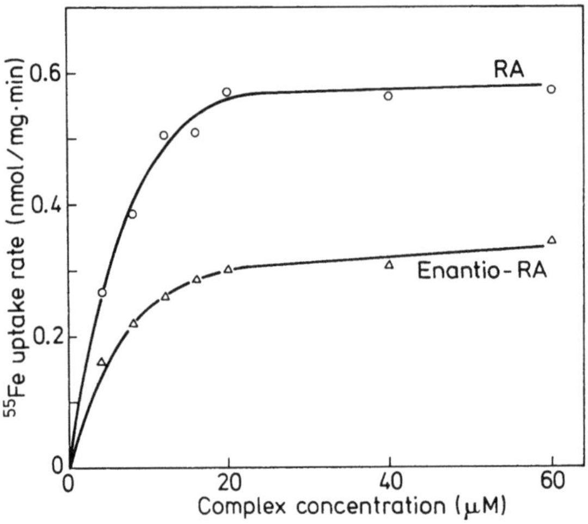

Fig. 32. The iron(III) uptake in *Rhodotorula pilimanae* mediated by Δ-Fe(III) rhodotorulate and its mirror image Λ-Fe(III)-enantio-rhodotorulate

with synthetic enantio-RA[77] clearly demonstrate the preferential recognition of the naturally-occurring Δ-RA complex compared to the Λ-enantio RA complex [210] (Fig. 32).

A number of enteric bacteria, including *E. coli* K-12, have a specific outer membrane receptor for ferrichrome, although they do not produce this siderophore. The ferrichrome receptor (Ton A or fhu A), also serves as a common binding site for the phages T5, T1 and θ 80 and the colicin M. Both the phages and colicin M kill the cells [65]. It is therfore highly desirable to obtain detailed information about the degree of specificity of the ferrichrome recognition. Unfortunately, until now the existing data were scattered in several papers which did not deal with the question of stereospecificity.

There exists a whole series of natural ferrichrome homologs which exhibit additional functionalities of the cyclic hexapeptide backbone or of the hydroxamate groups. For example, ferrichrome, and ferrichrysin which differ in the amino acid sequences of the backbone (Fig. 1, ferrichrome diagram: ferrichrysin: $R' = R'' = CH_2OH$, $R''' = H$, $R = CH_3$) are taken up by *E. coli* at similar rates. This is also true for a succinoyl derivative of ferrichrysin [209] and ferricrocin [211] (Fig. 1, ferrichrome diagram: $R' = R''' = H$, $R'' = CH_2OH$, $R = CH_3$). The antibiotic albomycin is able to compete with ferrichrome for the Ton A receptor site [64]. Albomycin contains in the cyclic hexapeptide a triseryl building block instead of a triglycyl peptide building block in the case of ferrichrome. In contrast, ferrichrome A (a ferrichrysin derivative with changes in the residues of the hydroxamate portion of the molecule) is reported to be growth inactive [212]. In another experiment Winkelmann and Braun showed that the uptake of Δ-*cis* enantio-ferrichrome is reduced by about 50% compared with the natural Λ-*cis* ferrichrome [213].

Briefly summarized, it appears that any changes at the hydroxamate complex "propeller" part of the molecule affect the uptake rate dramatically, whereas considerable modifications of the backbone do not show such pronounced effects. Again it may be concluded that the metal center and adjacent functionalities are the key for the recognition by the receptor. In this light it seems possible that the ferrichrome competing phages and colicin M should possess some structural features similar to the hydroxamate functionality of the ferrichrome molecule.

Our current knowledge of the stereospecificity of siderophore uptake remains still very fragmentary. Therefore the formulation of a generally accepted rule is not possible. However one trend conforms to all the data at hand so far: it seems that the predominantly recognized part of the molecule is the iron "octahedral" coordination site and its adjacent functionalities.

IX Acknowledgments

The siderophore research and iron chelation projects of this laboratory are supported by NIH Grants AI 11744 and HL 24775. G.M. thanks the Deutsche Forschungs Gemeinschaft for a fellowship. We are pleased to thank Ms. June Smith and several members of this research group for their editorial assistance.

X References

1. Twort, F. W., Ingram, G. L. Y.: Proc. Roy. Soc. Ser. B. Biol. Sci. *84*, 517 (1912)
2. Francis, J., Madinaveitia, J., Macturk, H. M., Snow, G. A.: Nature *163*, 365 (1949)
3. Francis, J., Macturk, H. M., Madinaveitia, J., Snow, G. A.: Biochem. J. *55*, 596 (1953)
4. Snow, G. A.: J. Chem. Soc., 4080 (1954)
5. For a complete chronicle of the mycobactins see: Snow, G. A.: Bacteriol. Rev. *34*, 99 (1970)
6. Horowitz, N. H., Charlang, G., Horn, G., Williams, N. P.: J. Bacteriol. *127*, 135 (1976)
7. See, for example, Glynn, A. A.: Sym. Soc. Gen. Microbiol. *22* 75 (1972); Rogers, H. J.: Infect. Immun. *7*, 445 (1973)
8. Schroth, M. N., Hancock, J. G.: Science *216*, 1376 (1982)
9. Olsen, R. A., Clark, R. B., Bennett, J. H.: Am. Scientist *69*, 378 (1981)
10. Anderson, W. F., Hiller, M. C. (Ed.): Development of Iron Chelators for Clinical Use, Bethesda, Maryland 1975
11. Anderson, W. F.: ACS Symp. Series *140*, 251, 1980
12. Raymond, K. N., Smith, W. L.: Structure and Bonding *43*, 159, 1981
13. Development of Iron Chelators for Clinical Use. Martell, A. E., Anderson, W. F., Badman, D. G., Eds., Elsevier/North-Holland, New York, 1981
14. Raymond, K. N., Pecoraro, V. L., Harris, W. R., Carrano, C. J. in: Environmental Migration of Long-Lived Radionuclides, IAEA, Vienna 1982, p. 571
15. Bulman, R. A.: Structure and Bonding *34*, 39, 1978
16. Raymond, K. N., Chung, T. D. Y., Pecoraro, V. L., Carrano, C. J. in: The Biochemistry and Physiology of Iron, Saltman, P., Hegenauer, J., Eds., Elsevier Biomedical, New York 1982, p. 649
17. Tidmarsh, G. F., Klebba, Ph. E., Rosenberg. L. L.: J. Inorg. Biochem. *18*, 161 (1983)
18. Neilands, J. B.: Ann. Rev. Nutr. *1*, 27 (1981)
19. Neilands, J. B. (Ed.): Microbial Iron Metabolism, Academic Press, New York 1974
20. Emery, T.: Metal Ions in Biological Systems *7*, 77 (1977)
21. Byers, B. R., Arcenaux, J. E. L. in: Microorganisms and Minerals, Weinberg, E. D., Ed., Marcel Dekker, New York 1977, p. 215
22. Lankfort, C. E.: CRC Crit. Rev. Microbiol. *2*, 273 (1973)
23. Weinberg, E. D.: Microbiol. Rev. *42*, 45 (1978)
24. Raymond, K. N., Harris, W. R., Carrano, C. J., Weitl, F. L.: ACS Symp. Ser. *140*, 313, 1980
25. Waid, J. S. in: Soil Biochemistry, Paul, E. A., McLaren, A. D., Eds., Marcel Dekker, New York 1974, p. 65
26. Neilands, J. B.: Structure and Bonding *1*, 59, 1966

27. Neilands, J. B. in: Inorganic Biochemistry, Eichhorn, G. L., Ed., Elsevier Sci. Publ. Co., Amsterdam 1973, p. 167
28. Neilands, J. B., Ratledge, C. in: Handbook of Microbiology, Vol. IV, Laskin, A. I., Lechevalier, H. A., Eds., CRC Press, Boca Raton, Florida 1982, p. 565
29. Neilands, J. B.: Ann. Rev. Biochem. *50*, 715 (1981)
30. Harris, W. R., Carrano, C. J., Raymond, K. N.: J. Am. Chem. Soc. *101*, 2722 (1979)
31. Adapa, S., Huber, P., Keller-Schierlein, W.: Helvetica Chimica Acta *65*, 1818 (1982)
32. Hough, E., Rogers, D.: Biochem. Biophys. Res. Commun. *57*, 73 (1974)
33. Neilands, J. B., Peterson, T., Leong, S. A.: ACS Symp. Ser. *140*, 263, 1980
34. Teintze, M., Hossain, M. B., Barnes, C. I., Leong, J., van der Helm, D.: Biochemistry *20*, 6446 (1981)
35. Teintze, M., Leong, J.: ibid. *20*, 6457 (1981)
36. Carrano, C. J., Copper, S. R., Raymond, K. N.: J. Am. Chem. Soc. *101*, 599 (1979)
37. Bell, S. J., Friedman, S. A., Leong, J.: Antimicrobial Agents and Chemotherapy *15*, 384 (1979)
38. Latimer, W. M.: Oxidation Potentials, Prentice Hall, Englewood Cliffs, New Jersey 1952
39. Philson, S. B., Llinás, M. J.: Biol. Chem. *257*, 8086 (1982)
39. Philson, S. B., Llinás, M. J.: Biol. Chem. *257*, 8086 (1982)
40. Philson, S. B., Llinás, M. J.: ibid. *257*, 8081 (1982)
41. Hulcher, F. H.: Biochemistry *21*, 4491 (1982)
42. Emery, T.: Biochim. Biophys. Acta *629*, 382 (1980)
43. Frederick, C. B., Bentley, M. S., Shive, W.: Biochemistry *20*, 2436 (1981)
44. Yancey, R. J., Finkelstein, R. A.: Infection Immunity *32*, 600 (1981)
45. Cox, C. D., Graham, R.: J. Bacteriol. *137*, 357 (1979)
46. McCullough, W. G., Merkal, R. S.: ibid. *137*, 243 (1979)
47. Takemoto, T., Nomoto, K., Fushiya, S., Ouchi, R., Kusano, G., Hikino, H., Takagi, S., Matsuura, Y., Kakudo, M.: Proc. Japan. Acad. *54 Ser. B.*, 469, 1978
48. Sugiura, Y., Tanaka, H.: J. Am. Chem. Soc. *103*, 6979 (1981)
49. Fushiya, S., Sato, Y., Nozoe, S.: Tetrahedron Lett. *21*, 3071 (1980)
50. Nomoto, K., Yoshioka, H., Arima, M., Fushiya, S., Takagi, S., Takemoto, T.: Chimia *35*, 249 (1981)
51. Budé^{n}ínský, M., Budzikiewicz, H., Procházka, Ź., Ripperger, H., Roemer, A., Scholz, G., Schreiber K.: Phytochemistry *19*, 2295 (1980)
52. Nienhuis, A. W., Anderson, W. F. in: Development of Iron Chelators for Clinical Use, Anderson, W. F., Hiller, M. C., Eds., Elsevier/North-Holland, New York 1981
53. Ellis, J. T., Schulman, I., Smith, C. H.: Am. J. Pathol. *30*, 287 (1954)
54. Walker, R. J., Williams, R. in: Iron in Biochemistry and Medicine, Jacobs, A., Worwood, M., Eds., Academic Press, New York 1974, p. 596
55. Wohler, F.: Acta Haematol. *30*, 65 (1963)
56. Barry, M., Flynn, D. M., Letsky, E. A., Risdon, R. A.: Br. Med. J. *1*, 16 (1974)
57. Pippard, M. J., Callender, S. T., Letsky, E. A., Weatherall, D. J.: Lancet, 1178 (1978)
58. Hallberg, L., Hedenberg, L.: J. Haematol. *2*, 67 (1965)
59. Morgan, E. H.: Biochim. Biophys. Acta *244*, 103 (1971)
60. Waxman, H. S., Brown, E. B.: Progr. Hematol. *6*, 338 (1969)
61. Maehr, H.: Pure Appl. Chem. *28*, 603 (1971)
62. Emery, T.: Adv. Enzymol. Relat. Areas Mol. Biol. *35*, 135 (1971)
63. Zähner, H., Diddens, H., Keller-Schierlein, W., Naegeli, H. U.: Jap. J. Antibiotics *30*, 5 (1977)
64. Hartmann, A., Fiedler, H.-P., Braun, V.: Eur. J. Biochem. *99*, 517 (1979)
65. Braun, V., Hantke, K. in: Organization of Procariotic Cell Membranes, Vol. II, Ghosh, B. K., Ed., CRC Press, Boca Ratan, Florida 1981, pp. 1–73
66. Fiss, E. H., Stanley-Samuelson, P., Neilands, J. B.: Biochemistry *21*, 4517 (1982)
67. Bickel, H., Hall, E., Keller-Schierlein, W., Prelog, V., Vischet, E., Wettstein, A.: Helvet. Chimica Acta *43*, 2129 (1960)
68. Atkins, C. L., Neilands, J. B.: Biochemistry *7*, 3734 (1968)
69. Diekmann, H.: Arch. Mikrobiol. *73*, 65 (1970)
70. Keller-Schierlein, W., Diekmann, H.: Helv. Chim. Acta *53*, 2035 (1970)
71. Emery, T.: Biochemistry *4*, 1410 (1965)
72. Sayer, J. M., Emergy, T.: ibid. *7*, 184 (1968)

73. Diekmann, H.: Angew. Chem. *7*, 551 (1968)
74. Widmer, J., Keller-Schierlein, W.: Helv. Chim. Acta *57*, 1904 (1974)
75. Keller-Schierlein, W.: ibid. *52*, 603 (1969)
76. Naegeli, H. U., Keller-Schierlein, W.: ibid. *61*, 2088 (1978)
77. Isowa, Y., Takashima, T., Ohmori, M., Kurita, H., Sato, M., Mori, K.: Bull. Chem. Soc. Jap. *45*, 1461 and 1464 (1972)
78. Isowa, Y., Takashima, T., Ohmori, M., Kurita, H., Sato, M., Mori, K.: ibid. *45*, 1467 (1972)
79. Isowa, Y., Ohmori, M., Kurita, H.: ibid. *47*, 215 (1974)
80. Maurer, P. J., Miller, M. J.: J. Am. Chem. Soc. *104*, 3096 (1982)
81. Macham, L. P., Ratledge, C., Norton, J. C.: Infect. Immun. *12*, 1242 (1975)
82. Macham, L. P., Ratledge, C.: J. Gen. Microbiol. *89*, 379 (1975)
83. Lee, B. H., Miller, M. J.: J. Org. Chem. *48*, 24 (1983)
84. Maurer, P. J., Miller, M. J.: J. Am. Chem. Soc. *105*, 240 (1983)
85. Atkins, C. L.: Thesis, Univ. of California, Berkeley 1970
86. Smith, W. L., Raymond, K. N.: J. Am. Chem. Soc. *102*, 1252 (1980)
87. Rodgers, S. J., Raymond, K. N.: J. Med. Chem. *26*, 439 (1983)
88. Olson, R. K., Ramasamy, K., Emery, T.: J. Am. Chem. Soc. in press
89. Carrano, C. J., Raymond, K. N.: ibid. *101*, 5401 (1979)
90. Jacobs, A., White, G. P., Tait, G. P.: Biochem. Biophys. Res. Comm. *74*, 1626 (1977)
91. Bergeron, R. J., Kline, S. G.: J. Am. Chem. Soc. *104*, 4489 (1982)
92. Weitl, F. L., Raymond, K. N.: ibid. *101*, 2728 (1979). Note in particular references for DCC mediated condensations of amino acids with unprotected catechols. According to this paper enterobactin synthesis (Ref. 98) was performed with protected catechols under the reaction conditions described
93. Pollack, J. R., Neilands, J. B.: Biochem. Biophys. Res. Comm. *38*, 989 (1971)
94. O'Brien, I. G., Gibson, F.: Biochem. Biophys. Acta *215*, 393 (1970)
95. O'Brien, I. G., Cox, G. B., Gibson, F.: ibid. *237*, 537 (1971)
96. Bryce, G. F., Weller, R., Brot, N.: Biochem. Biophys. Res. Comm. *42*, 871 (1971)
97. Guterman, S. K., Morris, P. M., Tannenberg, W. J. K.: Gen. Pharmacol. *9*, 123 (1978)
98. Corey, E. J., Battacharyya, S.: Tetrahedron Lett. *45*, 3919 (1977)
99. Rastetter, W. H., Erickson, T. J., Venuti, M. C.: J. Org. Chem. *46*, 3579 (1981)
100. Shanzer, A., Libman, J.: J. Chem. Soc., Chem. Commun., 846 (1983)
101. Harris, W. R., Weitl, F. L., Raymond, K. N.: ibid. 176 (1979)
102. Pecoraro, V. L., Weitl, F. L., Raymond, K. N.: J. Am. Chem. Soc. *103*, 5133 (1981)
103. Weitl, F. L., Harris, W. R., Raymond, K. N.: J. Med. Chem. *22*, 1281 (1979)
104. Venuti, M. C., Rastetter, W. H., Neilands, J. B.: ibid. *22*, 123 (1979)
105. Lodge, G. S., Gaines, C. G., Arcenaux, J. E. L., Byers, B. R.: Biochem. Biophys. Res. Commun. *97*, 1291 (1980)
106. Heidinger, S., Braun, V., Pecoraro, V. L., Raymond, K. N.: J. Bacteriol. *153*, 109 (1983)
107. Collins, D. J., Lewis, C., Swan, J. M.: Aust. J. Chem. *27*, 2593 (1974)
108. Corey, E. J., Hurt, S. D.: Tetrahedron Lett. *45*, 3923 (1977)
109. Collins, D. J., Lewis, C., Swan, J. M.: Aust. J. Chem. *28*, 673 (1975)
110. Hollifield, W. C., Jr., Neilands, J. B.: Biochemistry *17*, 1922 (1978)
111. Tait, G. H.: Biochem. J. *146*, 191 (1975)
112. Peterson, T., Falk, K.-E., Leong, S. A., Klein, M. P., Neilands, J. B.: J. Am. Chem. Soc. *102*, 7715 (1980)
113. Harris, W. R., Raymond, K. N.: ibid. *101*, 6534 (1979)
114. Grady, R. W., Graziano, J. H., Akers, K. A., Cerami, A. J.: Pharmacol. Exp. Ther. *196*, 478 (1976)
115. Weitl, F. L., Raymond, K. N., Durbin, P. W.: J. Med. Chem. *24*, 203 (1981)
116. Goldstein, A., Aronow, L., Kalman, S. K.: Principles of Drug Action, Wiley, New York, 1974
117. Weitl, F. L., Raymond, K. N.: J. Org. Chem. *46*, 5234 (1981)
118. Raymond, K. N., Kappel, M. J., Pecoraro, V. L., Harris, W. R., Carrano, C. J., Weitl, F. L., Durbin, P. W. in: Actinides in Perspective, Edelstein, N. M., Ed., Pergamon Press, Oxford and New York 1982, pp. 491–507
119. Sofen, S. R., Cooper, S. R., Raymond, K. N.: Inorg. Chem. *18*, 1611 (1979)

120. Weitl, F. L., Raymond, K. N.: J. Am. Chem. Soc. *102*, 2289 (1980)
121. Kappel, M. J., Raymond, K. N.: submitted for publication in J. Am. Chem. Soc.
122. Harris, W. R., Carrano, C. J., Cooper, S. R., Sofen, S. R., Avdeef, A., McArdle, J. V., Raymond, K. N.: J. Am. Chem. Soc. *101*, 6097 (1979)
123. Harris, W. R., Raymond, K. N., Weitl, F. L.: ibid. *103*, 2667 (1981)
124. Harris, W. R., Carrano, C. J., Raymond, K. N.: ibid. *101*, 2213 (1979)
125. Anderegg, G., L'Eplattenier, F., Schwarzenbach, G.: Helv. Chim. Acta *46*, 1409 (1963)
126. Kappel, M. J., Raymond, K. N.: Inorg. Chem. *21*, 3437 (1982)
127. Birus, M., Bradić, Z., Kugunžić, N., Pribanić, M.: Inorg. Chim. Acta *78*, 87 (1983)
128. Monzyk, B., Crumbliss, A. L.: J. Am. Chem. Soc. *104*, 4921 (1982)
129. Pecoraro, V. L., Wong, G. B., Kent, T. A., Huynh, B. H., Raymond, K. N.: ibid. *105*, 4617 (1983)
130. Pecoraro, V. L., Harris, W. R., Wong, G. B., Carrano, C. J., Raymond, K. N.: ibid. *105*, 4623 (1983)
131. Stover, B. J., Brunger, F. W.: Radiat. Res. *33*, 381 (1968)
132. Welch, M. J., Welch, T. J. in: Radiopharmaceuticals, Supramanian, G., Roads, P. G., Copper, D. J., Sodd, R. C., Eds., Soc. of Nuclear Medicine, New York 1975, pp. 73–79
133. Clausen, J., Edeling, C. J., Fogh, J.: Cancer Res. *34*, 1931 (1974)
134. Noujain, A. A., Lentle, B. C., Hill, J. R., Terner, U. K., Wong, H.: Int. J. Nucl. Med. Biol. *6*, 193 (1979)
135. Shaiman, R. D.: Acta Crystallogr., Sect. H. *H 32*, 751 (1976)
136. Taylor, G. N., Williams, J. L., Roberts, L., Albertan, D. R., Shabestari, L.: Health Phys. *27*, 285 (1974)
137. Edwards, G. L., Hayes, R. L.: J. Nucl. Med. *10*, 103 (1969)
138. Harper, P. V.: Int. J. Appl. Rad. Isotopes *28*, 5 (1977)
139. Gelb, M. H., Harris, D. C.: Arch. Biochem. Biophys. *200*, 439 (1980)
140. Woodworth, R. C., Morallee, K. G., Williams, R. J. P.: Biochemistry *9*, 839 (1970)
141. Pecoraro, V. L., Wong, G. B., Raymond, K. N.: Inorg. Chem. *21*, 2209 (1982)
142. Lehninger, A. L.: Biochemie, Verlag Chemie Weinheim, 1979
143. Robinson, J. P., McArdle, J. V.: J. Inorg. Nucl. Chem. *43*, 1951 (1981)
144. Rosenberg, H., Young, I. Y. in: Microbial Iron Metabolism, Neilands, J. B., Ed., Academic Press, New York 1974, pp. 67–82
145. Langman, L., Young, I. G., Trost, G., Rosenberg, H., Gibson, F.: J. Bacteriol. *112*, 1142 (1972)
146. Bryce, G. F., Brot, N.: Biochemistry *11*, 1708 (1972)
a) Ball, E. G., Chem, T. T.: J. Biol. Chem. *102*, 691 (1933)
147. Mentasti, E., Pelizetti, E., Saini, G.: J. Chem. Soc. Dalton Trans., 2609 (1973); J. Inorg. Nucl. Chem. *38*, 785 (1976)
148. Hider, R. C., Mohd-Nor, A. R., Silver, J., Morrison, I. E. G., Rees, L. V. C.: J. Chem. Soc. Dalton Trans., 609 (1981)
149. Howlin, B., Hider, R. C., Silver, J.: ibid. 1433 (1982)
150. Kipton, H., Powell, J., Taylor, M. C.: Aust. J. Chem. *35*, 739 (1982)
151. Hider, R. C., Silver, J., Neilands, J. B., Morrison, I. E. G., Rees, L. V. G.: FEBS Lett. *102*, 325 (1979)
152. Hamed, M. Y., Hider, R. C., Silver, J.: Inorg. Chim. Acta *66*, 13 (1982)
153. Matzanke, B. F., Huynh, B. H., Raymond, K. N.: Paper in preparation
154. Harris, W. R., Pecoraro, V. L.: Biochemistry *22*, 292 (1983)
155. Tufano, T. P., Pecoraro, V. L., Raymond, K. N.: Biochim. Biophys. Acta *668*, 420 (1981)
156. Monzyk, K. B., Crumbliss, A. L.: Inorg. Chim. Acta *55*, L 5 (1981)
157. Kazmi, S. A., McArdle, S. V.: Inorg. Biochem. *15*, 153 (1981)
158. Monzyk, K. B., Crumbliss, A. L.: J. Am. Chem. Soc. *101*, 6203 (1979)
159. Biruš, M., Kujundzić, N., Pribanić, M.: Inorg. Chim, Acta *55*, 65 (1980)
160. Kazmi, S. A., McArdle, J. V.: J. Inorg. Nucl. Chem. *43*, 3031 (1981)
161. Lovenberg, W., Buchanan, B. B., Rabinowtiz, J. C.: J. Biol. Chem. *236*, 3899 (1963)
162. Basolo, F., Pearson, R. G. in: Mechanisms of Inorganic Reactions, 2nd Ed., John Wiley & Sons, Inc., New York 1967, p. 152
163. Emery, T., Hoffer, P. B.: J. Nucl. Med. *21*, 935 (1980)
164. Tufano, T. P.: Ph. D. Dissertation, Univ. of California, Berkeley, California 1982

165. Tufano, T. P., Raymond, K. N.: J. Am. Chem. Soc. *103*, 6617 (1981)
166. Raymond, K. N., Tufano, T. P. in: The Biological Chemistry of Iron, Dunford, H. B., Dolphin, D., Raymond, K. N., Seiker, L., Eds., D. Reidel Publishing, Boston, London 1982, pp. 85–105
167. Olson, D. C., Margerum, D. W.: J. Am. Chem. Soc. *85*, 297 (1963)
168. Wickmann, H. H., Klein, M. P., Shirley, D. A.: J. Chem. Phys. *42*, 2113 (1965)
169. Dowsing, R. D., Gibson, J. F.: ibid. *50*, 294 (1968)
170. Wickmann, H. H., Klein, M. P., Shirley, D. A.: Phys. Rev. *152*, 345 (1966)
171. Bock, J. L., Lang, G.: Biochim. Biophys. Acta *264*, 245 (1972)
172. Spartalian, K., Osterhuis, W. T., Neilands, J. B.: J. Chem. Phys. *62*, 3538 (1975)
173. Ehrenberg, A.: Nature *178*, 379 (1956)
174. Raymond, K. N., Abu-Dari, K., Sofen, S. R.: ACS Symp. Ser. *119*, 133, 1980
175. Leong, J., Raymond, K. N.: J. Am. Chem. Soc. *96*, 6628 (1974)
176. Leong, J., Raymond, K. N.: ibid. *97*, 293 (1975)
177. Müller, G., Barclay, S. J., Raymond, K. N.: Manuscript in preparation
178. Isied, S. S., Kuo, G., Raymond, K. N.: J. Am. Chem. Soc. *98*, 1763 (1976)
179. Leong, J., Bell, S. J.: Inorg. Chem. *17*, 1886 (1978)
180. Hossain, M. B., Eng-Wilmot, D. L., Loghry, R. A., van der Helm, D.: J. Am. Chem. Soc. *102*, 5766 (1980)
181. van der Helm, D., Baker, J. R., Eng-Wilmot, D. L., Hossain, M. B., Loghry, R. A.: ibid. *102*, 4224 (1980)
182. Raymond, K. N. in: Advances in Chemistry Series No. 162, Bioinorganic Chemistry-II, J. Am. Chem. Soc., 1977, p. 33
183. Abu-Dari, K., Ekstrand, J. D., Freyberg, D. P., Raymond, K. N.: Inorg. Chem. *18*, 108 (1979)
184. Leong, J., Raymond, K. N.: J. Am. Chem. Soc. *96*, 1757 (1974)
185. Zalkin, A., Forrester, J. D., Templeton, D. H.: ibid. *88*, 1810 (1966)
186. Norrestam, R., Stensland, B., Brándén, C. J.: J. Mol. Biol. *99*, 501 (1975)
187. van der Helm, D., Poling, M.: J. Am. Chem. Soc. *98*, 82 (1976)
188. Carrano, C. J., Raymond, K. N.: ibid. *100*, 5371 (1978)
189. Barclay, S. J., Huynh, B. H., Raymond, K. N.: Paper in preparation
190. Barclay, S. J., Riley, P. E., Raymond, K. N.: J. Am. Chem. Soc. *104*, 6802 (1982)
191. White, D. L., Scarrow, R. C., Raymond, K. N.: Paper in preparation
192. Abu-Dari, K., Raymond, K. N.: J. Am. Chem. Soc. *99*, 2003 (1977)
193. Klebba, P. E., McIntosh, M., Neilands, J. B.: J. Bacteriol. *149*, 880 (1982)
194. Leong, J., Neilands, J. B., Raymond, K. N.: Biochem. Biophys. Res. Commun. *60*, 1066 (1974)
195. Arcenaux, J. E. L., Davis, W. B., Downer, D. N., Haydon, A. H., Byers, B. R.: J. Bacteriol. *115*, 919 (1973)
196. Leong, J., Neilands, J. B.: ibid. *126*, 823 (1976)
197. Brown, K. A., Ratledge, C.: FEBS Lett. *53*, 262 (1975)
198. Arcenaux, J. E. L., Byers, B. R.: J. Bacteriol. *141*, 715 (1980)
199. Gaines, C. G., Lodge, J. S., Arcenaux, J. E. L., Byers, B. R.: ibid. *148*, 527 (1981)
200. Straka, J. G., Emergy, T.: Biochim. Biophys. Acta *569*, 277 (1979)
201. Lodge, J. S., Gaines, C. G., Arcenaux, J. E. L., Byers, B. R.: J. Bacteriol. *149*, 771 (1982)
202. Bauminger, E. R., Cohen, S. G., Dickson, D. P. E., Levy, A., Ofer, S., Yariv, J.: Biochim. Biophys. Acta *623*, 237 (1980)
203. McIntosh, M. A., Earhart, C. F.: Biochem. Biophys. Res. Commun. *70*, 315 (1976)
204. Braun, V., Hancock, R. E. W., Hantke, K., Hartmann, A.: J. Supramol. Struct. *5*, 37 (1976)
205. Coulton, W.: Biochim. Biophys. Acta *717*, 154 (1982)
206. Ernst, J. F., Winkelmann, G.: ibid. *500*, 27 (1977)
207. Müller, G., Winkelmann, G.: FEMS Microbiol. Lett. *10*, 327 (1981)
208. Neilands, J. B., Erickson, J., Rastetter, W. H.: J. Biol. Chem. *256*, 3831 (1981)
209. Schneider, R., Hartmann, A., Braun, V.: FEMS Microbiol. Lett. *11*, 115 (1981)
210. Müller, G., Raymond, K. N.: Manuscript in preparation
211. Coulton, J. W., Naegeli, H.-U., Braun, V.: Eur. J. Biochem. *99*, 39 (1979)
212. Wayne, R., Frick, K., Neilands, J. B.: J. Bacteriol. *126*, 7 (1976)
213. Winkelmann, G., Braun, V.: FEMS Microbiol. Lett. *11*, 237 (1981)
214. Wong, G. B., Kappel, M. J., Raymond, K. N., Matzanke, B., Winkelmann, G.: J. Am. Chem. Soc. *105*, 810 (1983)

215. Schwarzenbach, G., Schwarzenbach, K.: Helv. Chim. Acta *46*, 1390 (1963)
216. Martell, A. E., Smith, R. M.: Critical Stability Constants, Plenum Press, New York 1977
217. Cooper, S. R., McArdle, J. V., Raymond, K. N.: Proc. Natl. Acad. Sci. USA *75*, 3551 (1978)
218. Anderson, B. F., Buckingham, D. A., Robertson, G. B., Webb, J., Murray, K. S., Clark, D. E.: Nature *262*, 772 (1976)
219. Letkeman, P., Martell, A. E., Motekaitis, R. J.: Coord. Chem. *10*, 47 (1980)
220. Harris, W. R., Martell, A. E.: Inorg. Chem. *15*, 713 (1976)
221. Chasteen, N. D.: Coord. Chem. Rev. *22*, 1 (1977)
222. Aasa, R., Malmström, B. G., Saltman, P., Vanngard, T.: Biochim. Biophys. Acta *75*, 203 (1963)
223. Emery, T.: Biochemistry *5*, 3694 (1966)
224. Ecker, D. J., Passavant, C. W., Emery, T.: Biochim. Biophys. Acta *720*, 242 (1982)
225. Keller-Schierlein, W., Deér, A.: Helv. Chim. Acta *46*, 1907 (1963)
226. Rastetter, W. H., Erickson, T. J., Venuti, M. C.: J. Org. Chem. *46*, 3579 (1981)
227. Rogers, H. J., Synge, C., Kimber, B., Bayley, P. M.: Biochim. Biophys. Acta *497*, 548 (1977)

Valence-Bond Isomers of Aromatic Compounds

Yoshiro Kobayashi[1] and Itsumaro Kumadaki[2]

1 Tokyo College of Pharmacy 1432-1, Horinouchi, Hachioji, Tokyo, 192-03, Japan
2 Faculty of Pharmaceutical Sciences, Setsunan University, 45-1, Nagao-Toge-cho, Hirakata, Osaka, 573-01, Japan

Table of Contents

After a brief survey of the history of valence-bond isomers of aromatic compounds, new syntheses and the reactions of these isomers reported in the last decade are reviewed. In the second chapter, the valence-bond isomers of homoaromatic compounds, especially benzene derivatives, are described and in the third chapter those of heterocyclic compounds. Photoreactions of perfluoroalkylated aromatic compounds afford valence-bond isomers in high yields. These isomers are very stable and useful for the synthesis of highly strained compounds. Therefore, the emphasis is put on the chemistry of trifluoromethylated benzvalenes, Dewar thiophenes, and Dewar pyrroles.

1 Introduction

In the course of the study of the structure of benzene, several structural formulas have been proposed for it. Well known structures are Kekulé's cyclohexatriene, Ladenburg's prismane, Thiele's partial bond structure, and Dewar's bicyclohexadiene. Later, Hückel proposed benzvalene as another structure of benzene. Since Kekulé's structure was gradually accepted as the structure of benzene, other structures are only of historical importance.

However, the progress of photochemistry suggested that some valence-bond isomers of benzene could play an important role in the isomerization of a substituted benzene. Some attempts were made to isolate such isomers in the 1960's. The first success was the isolation of Dewar benzene stabilized with tert-butyl groups by van Tamelen. This was the start of the isolation of many valence-bond isomers of aromatic compounds in the 1960's. Most of these isomers produced in the photoreaction are substituted by large substituents like a tert-butyl group.

In 1970, unsubstituted valence-bond isomers of benzene were synthesized on a preparative scale and their properties were fully investigated. Since the end of 1960, the chemistry of fluorine compounds was rapidly developed and the photolysis of perfluoro aromatic compound afforded valence-bond isomers in much higher yields than the corresponding hydrocarbon counterparts. These isomers were much more stable than the hydrocarbon analogs and could be used as starting materials for the synthesis of strained compounds.

In this review the progress of the chemistry of valence-bond isomers during and after the 1970's is discussed. For convenience, the chemistry of homoaromatic compounds is treated first followed by a discussion of heterocyclic compounds.

2 Valence-Bond Isomers of Homoaromatic Compounds

2.1 Syntheses of Dewar Benzenes

Baker and Rouvray [1] argued that the term "Dewar" benzene should be abandoned for the honour of Dewar, since he never believed that the bicyclohexadiene could be a structure of benzene. He only showed some possibilities of compounds by molecular model, which included Kekulé structure and others. The authors of this article don't agree with this proposal since we should pay more attention to Dewar's proposal of a highly strained compound. Furthermore, the term "Dewar benzene" has become too familiar to chemists to be abandoned. Thus, we use this term in this review.

Since Dewar's proposal about a hundred years had passed till van Tamelen [2] isolated a Dewar benzene obtained by irradiation of 1,2,4-tri-*tert*-butylbenzene:

(1)

Viehe and his coworkers obtained Dewar and benzvalene isomers by trimerization of *tert*-butylfluoroacetylene (2) [3]:

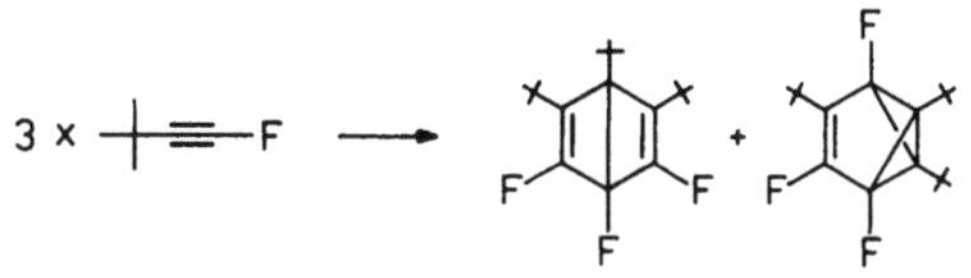

(2)

The synthesis of unsubstituted Dewar benzene was accomplished by van Tamelen (3) [4]. Thus, dihydrophthalic anhydride was photochemically isomerized to a [2.2.0]-ring system which was decarboxylated oxidatively to Dewar benzene. This compound has a higher stability than expected from the high strain of its ring system ($t_{1/2}$ = 37.2 h at 24.3°). This stability was later explained by the rule of the conservation of the orbital symmetry by Woodward and Hoffmann.

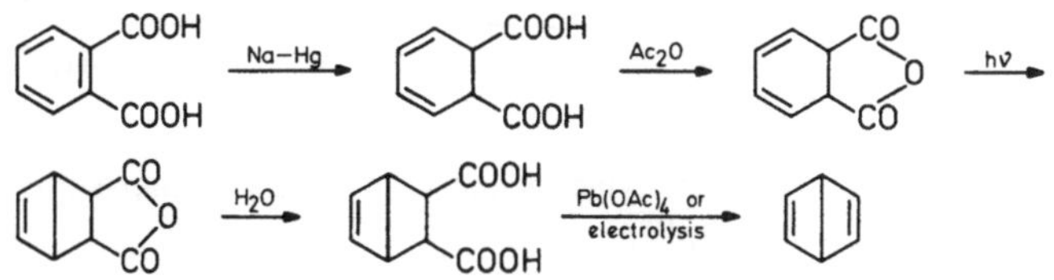

(3)

Irradiation of benzene with an oxygen lamp afforded Dewar benzene together with fulvene and benzvalene but the yield was very low so that this method has practically no preparative value [5].

Other preparative methods make use of cyclobutadiene intermediates which react with acetylene compounds. The first method was reported by Criegee and his coworker [6] who prepared tetramethylcyclobutadiene from the 1,2-diiodo derivative (4). In the second method, the cyclobutadiene irontricarbonyl complex is oxidized by Ce^{IV} to produce free cyclobutadiene (5) [7]. This reaction is widely used for the synthesis of cyclobutadiene.

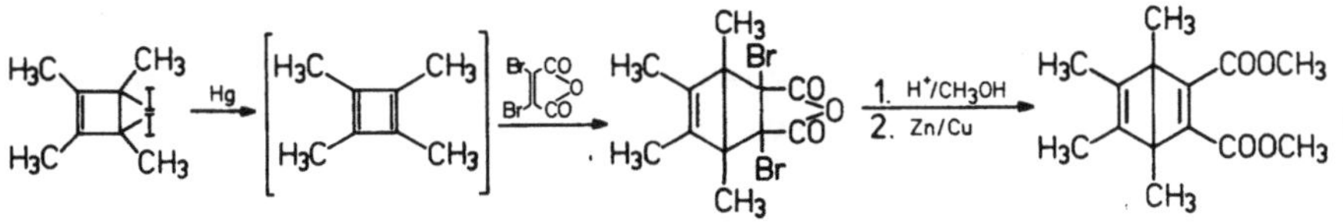

(4)

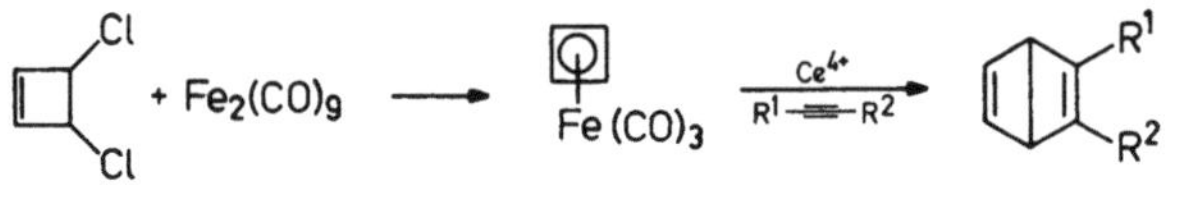

(5)

Furthermore, a more sterically hindered cyclobutadiene was converted into a Dewar benzene which is very stable and does not isomerized thermally to the corresponding benzene but to another Dewar isomer (6) [8].

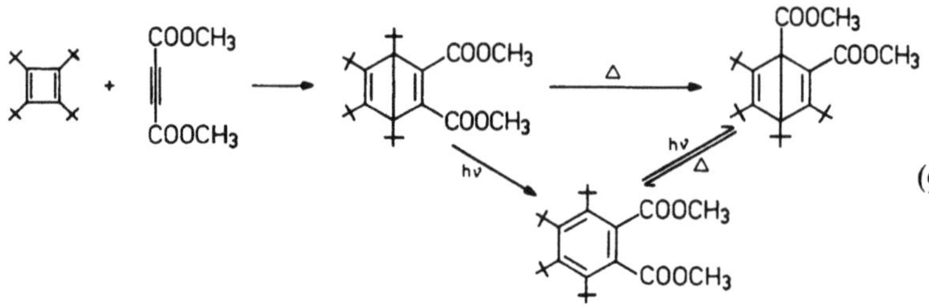

(9)

In this example, the repulsion between bulky substituents in a planar structure makes the Kekulé benzene less stable than the Dewar isomer.

The trimerization of alkynes to a Dewar benzene, which was found by Viehe's group, was developed by Schäfer's group using a Lewis acid as a catalyst. Thus, the treatment of 2-butyne with aluminum chloride gives hexamethyl Dewar benzene (7) [9]. Hexamethyl Dewar benzene is fairly stable and now commercially available.

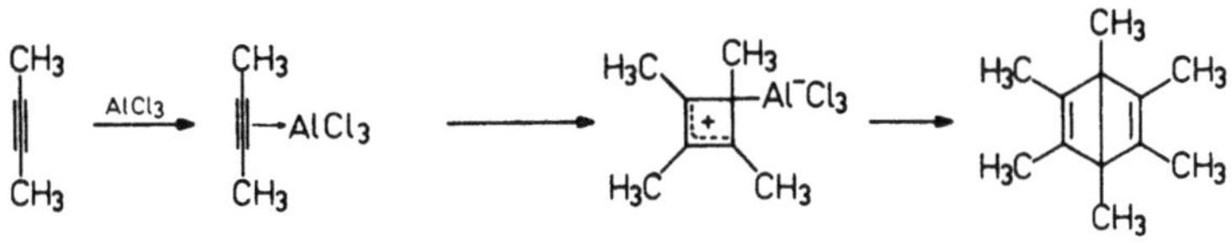

(7)

Another trimerization of diphenylacetylene catalyzed polymer-supported molybdenum was reported recently (8) [10].

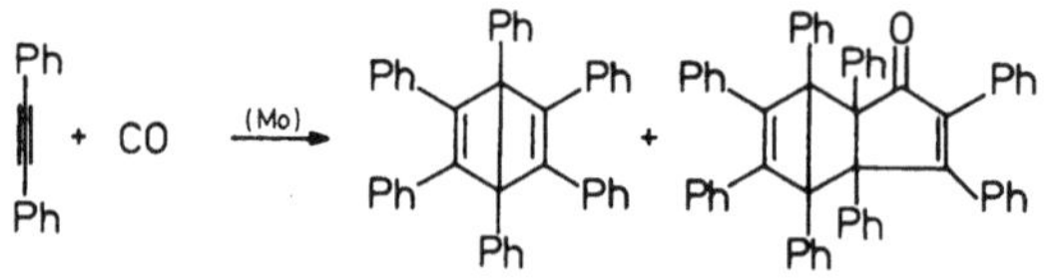

(8)

Photoisomerization of a sterically strained benzene to the Dewar isomer is exemplified by the photolysis of para-bridged benzene (9) [11]:

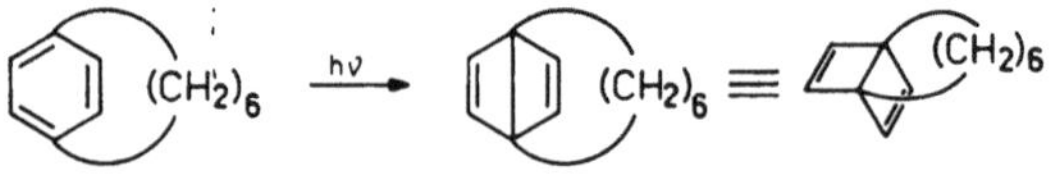

(9)

Another approach to a Dewar benzene was accomplished by ring contraction of 1,5-dimethyl-3,7-dithiabicyclo[3.3.0]octane (10) [12]:

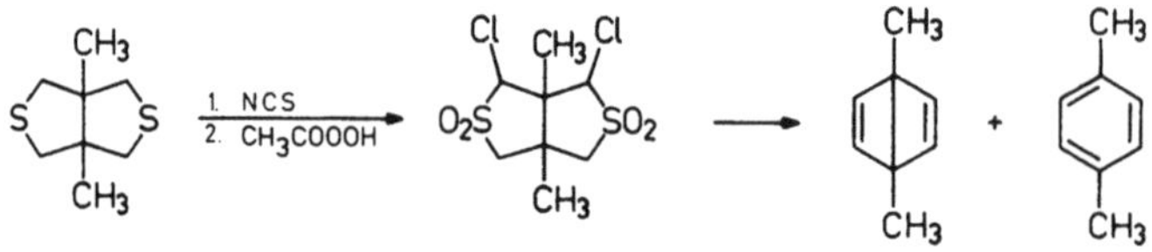

(10)

The synthesis of Dewar benzenes from hydrocarbons has been briefly surveyed. These examples show that photoreactions of strained benzenes give Dewar isomers, whereas other benzenes do not give Dewar isomers in preparative quantities so that the latter must be synthesized by cyclization reactions. In contrast to these hydrocarbons, the photoreaction of perfluorinated aromatic compounds gives the corresponding valence-bond isomers in good yields. Three of the most typical examples are described by Eqs. (11) to (13).

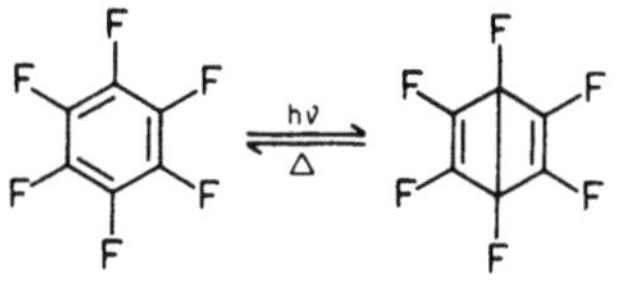

(11)

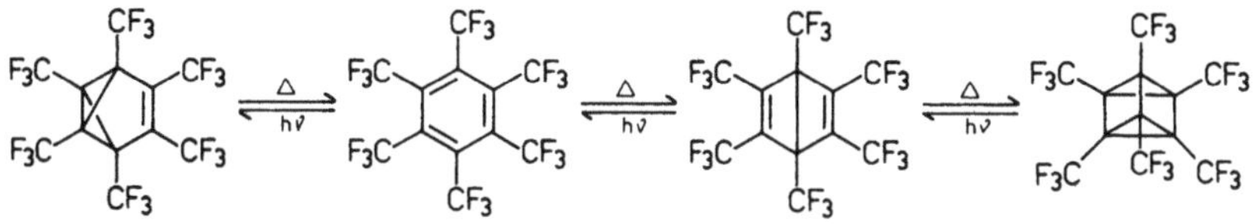

(12)

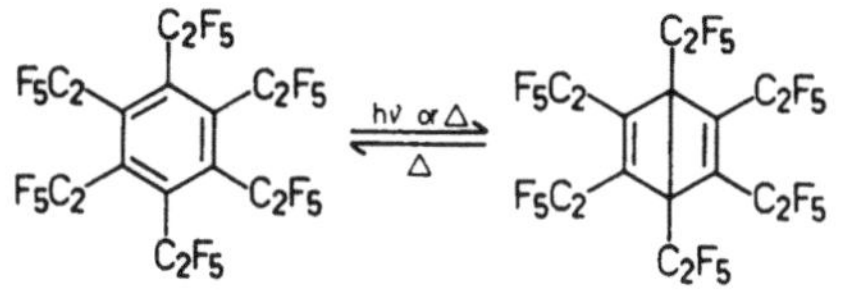

(13)

Hexafluorobenzene is converted into its Dewar isomer in good yield. This compound is stable in solution and thermally isomerizes to the corresponding benzene but it is explosive in the pure state (11) [13].

Hexakis(trifluoromethyl)benzene isomerizes to the Dewar isomer and further to the prismane on irradiation with short wavelength light while irradiation with light of longer wavelengths gives the benzvalene compound. This demonstrates that by appropriate choice of wavelength selective synthesis of valence-bond isomers may be achieved (12) [14,15]. The photolysis of hexakis(pentafluoroethyl)benzene gives only the Dewar isomer (13). Hexakis(pentafluoroethyl)benzene seems to be highly strained due to the repulsion between the bulky pentafluoroethyl groups; it probably becomes more stable when it is isomerized to a Dewar compound [16]. Actually, this was the first example where a Dewar isomer could be thermally prepared from the corresponding benzene though the thermolysis should be carried out in a flow system. The reason why the trifluoromethyl compound gives both Dewar and benzvalene isomers while the pentafluoroethyl compound yields only a Dewar isomer is not yet clear.

These Dewar isomers are stable at room temperature.

Photoisomerization of fluorinated aromatic compounds to Dewar isomers, which has widely been studied, is illustrated by the following equations.

In the photolysis of 1,2,4-trifluorobenzene yields 2,5- and 3,6-bound Dewar isomers whereas the 2,4-bound isomer is not formed (14) [17].

Yakobson et al.[18] described the photoisomerization of perfluoromesitylene to a Dewar isomer (15) while Haszeldine et al.[19] obtained prismane and perfluoro-(1,2,4-trimethylbenzene) besides the Dewar isomer. The latter compound is formed via another intermediate Dewar isomer which is indicated in Scheme 16 in square brackets.

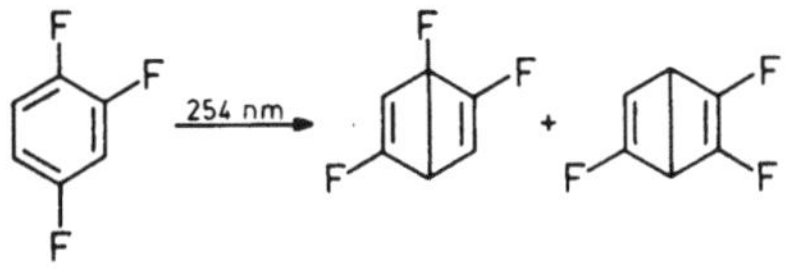

(14)

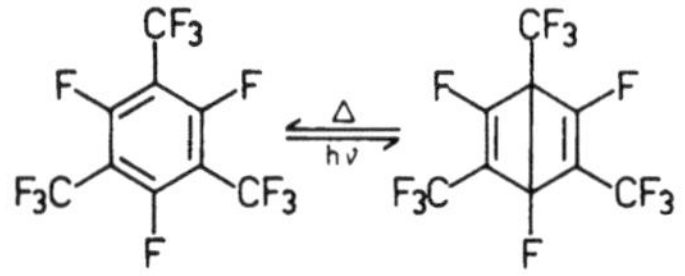

(15)

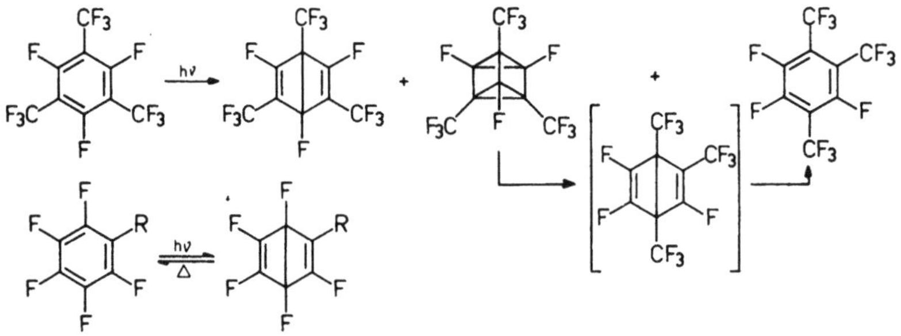

(16)

(17)

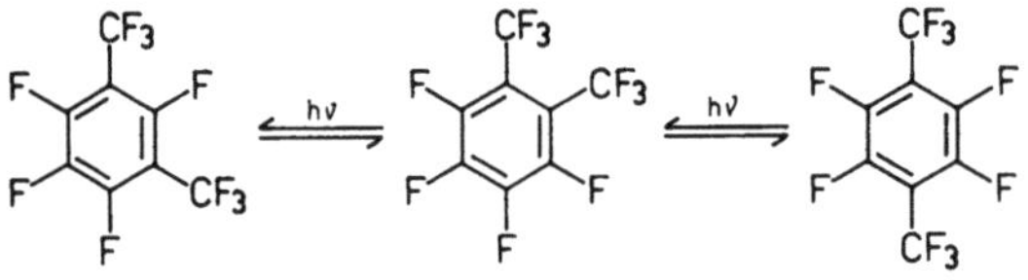

(18)

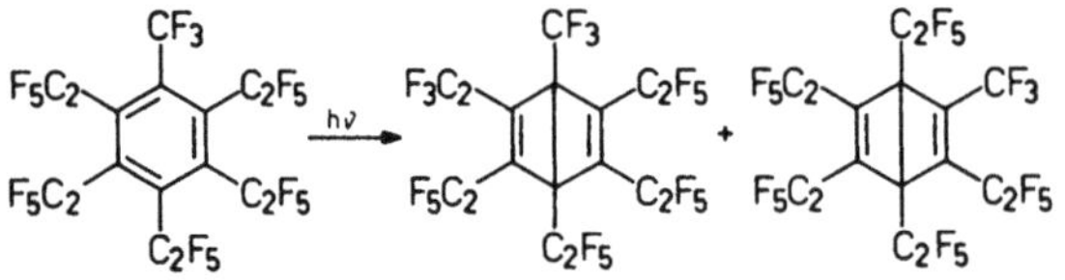

(19)

Pentafluorobenzene derivatives were also converted into Dewar isomers (17)[18] and perfluoroxylenes photoisomerized other isomers. The intermediates formed in these interconversions were confirmed to be Dewar isomers and trapped as the corresponding bromine adducts (18)[20].

Perfluoropentaethyltoluene was thermally isomerized to the two Dewar isomers shown in Eq. (19). Thermolysis of perfluorotetraethylxylenes also gives a mixture of Dewar isomers[21].

A special Dewar benzene is formed by the reaction of perfluoroisobutene with chloroform in the presence of cesium (I) fluoride (20) [22]:

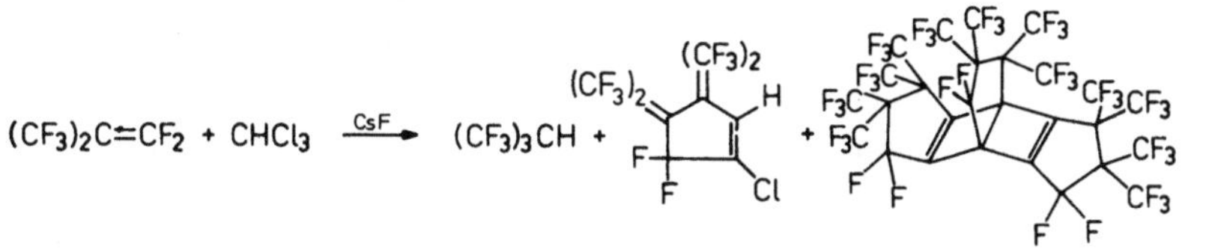

(20)

There are relatively few examples of Dewar isomers formed from condensed aromatic compounds. Only the photoisomerization of the *tert*-butylated compounds has been reported.

1,3,6,8-tetra-*tert*-butylnaphthalene is converted to the Dewar isomer in the presence of dianions (21) [23]:

(21)

A different product is however obtained when starting from tri-*tert*-butylnaphthalene [24]. The reaction is controlled by a chlorine atom in the 8-position. This is explained by the difference of the structures due to steric repulsions (22, 23).

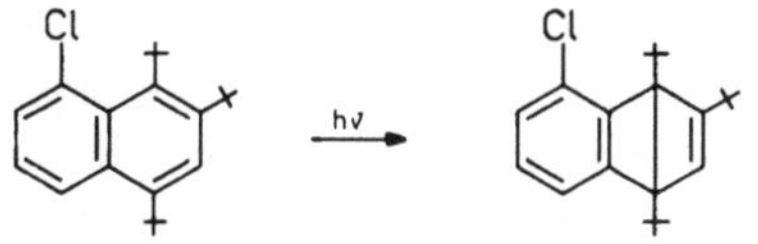

(22)

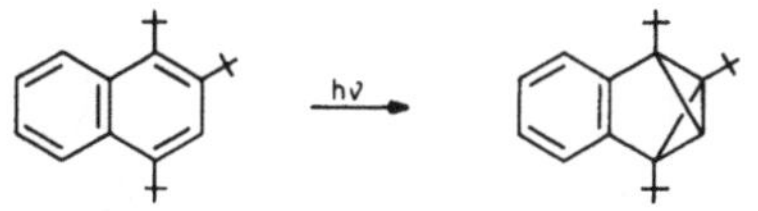

(23)

Photoisomerization of 9-*tert*-butylanthracene to the corresponding Dewar anthracene was studied with respect to the utilization of solar energy (24) [25].

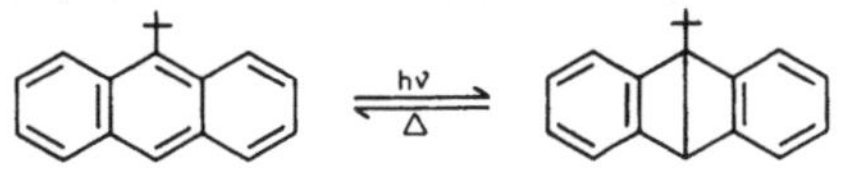

(24)

Other synthesis using cyclobutadiene intermediates are as follows (25, 26):

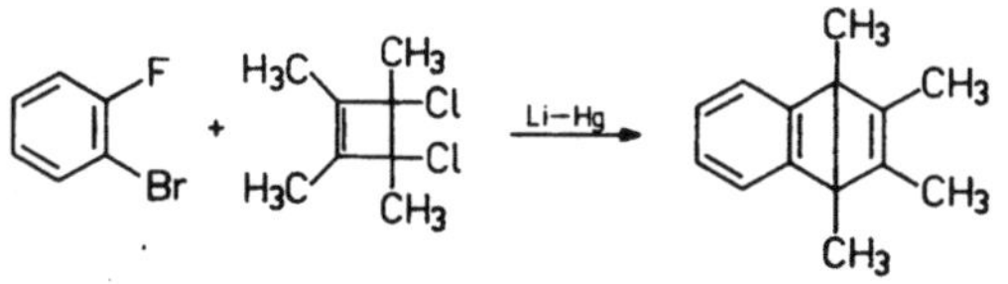

(25)

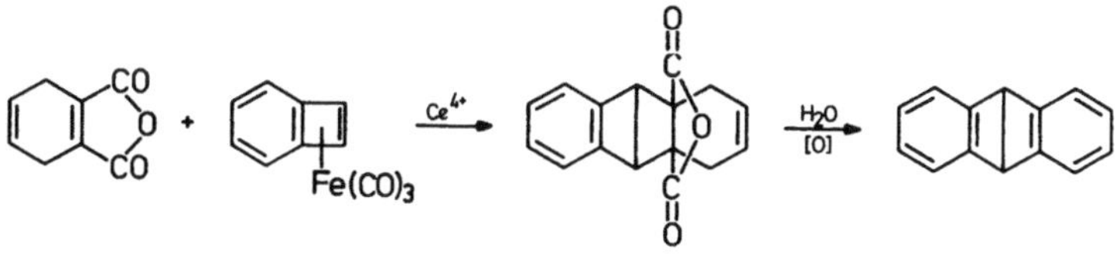

(26)

Thus, 1,2,3,4-tetramethyl Dewar naphthalene is formed by reaction of benzyne with tetramethylcyclobutadiene (25) [26]. Unsubstituted Dewar anthracene is obtained by a stepwise reaction (26) [27]. This compound violently isomerizes at 73 °C to anthracene. This shows that a *tert*-butyl group greatly affects the stability of the above mentioned 9-*tert*-butyl Dewar anthracene.

Another synthesis of a Dewar benzene part of a cross-condensed compound is based on the extrusion of sulfur (27) [28].

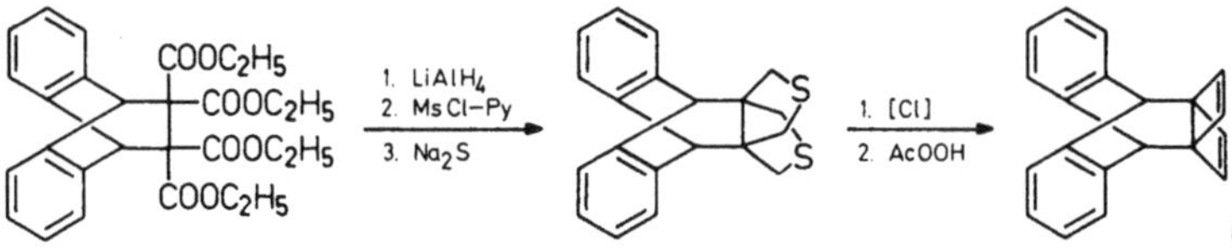

(27)

In the above-mentioned examples the Dewar compounds are isolated at room temperature. There are however also many reactions in which Dewar compounds are postulated as intermediates. Some of them are reported by Bryce-Smith and Gilbert [29].

2.2 Syntheses of Benzvalenes

Hückel suggested benzvalene structure as another possible isomer of benzene after the Dewar structure was proposed. The first synthesis of benzvalene isomer by trimerization of *tert*-butylfluoroacetylene was reported by Viehe's group [3] (2).

A careful reinvestigation of the photoreaction of tri-*tert*-butylbenzene showed that benzvalene is formed besides the corresponding Dewar benzene and prismane. This benzvalene is fairly stable (38) [30].

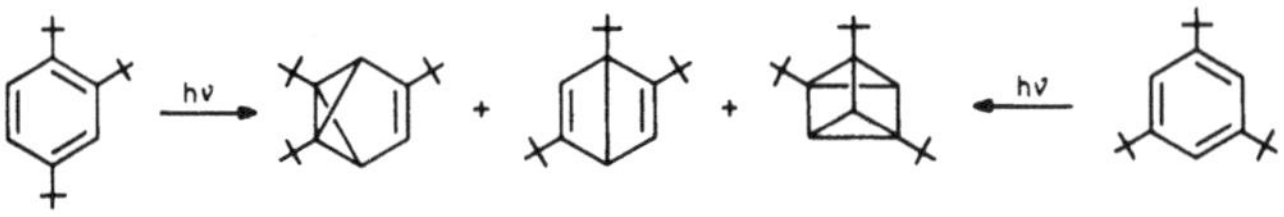

(28)

Wilzbach et al. observed that unsubstituted benzvalene is formed by photolysis of benzene. However, since the yield of benzvalene is very low this reaction had no preparative value (29) [31].

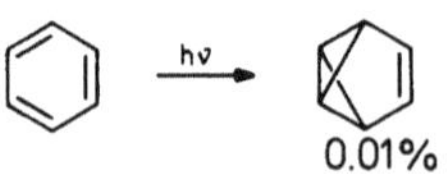

(29)

The synthesis of benzvalene on a preparative scale was achieved by Katz and his coworkers who treated lithium cyclopentadienide with methylene chloride and methyllithium and obtained benzvalene in good yield together with a small amount of benzene (30) [32].

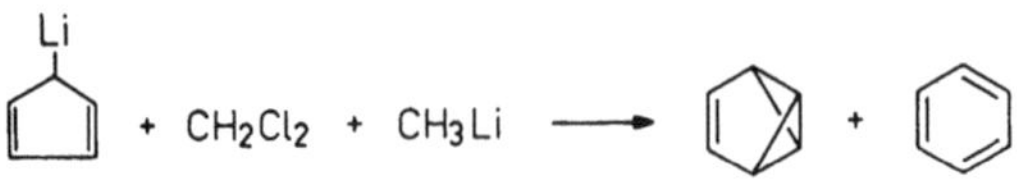

(30)

Pure benzvalene is explosive but in a solution it is fairly stable.

The mechanism of the formation of benzvalene was elucidated by Burger's group [33], who treated lithium cyclopentadienide and lithium indenide with dichlorodideuteriomethane and obtained 1-d-benzvalene, 2-d-benzobenzvalene, 1- and 2-d-naphthalene. Thus, a cyclopentadienylcarbene intermediate was postulated (32). Its rearrangement involving attack on the double bond (a) and migration of the phenyl (b) or vinyl group (c) leads to the products shown in scheme (32).

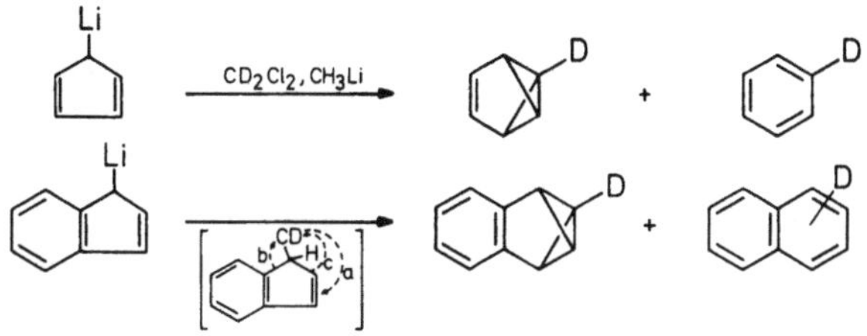

(31)
(32)

This mechanism was more extensively investigated by Burger's group [34].

The carbene produced from 5-methyl-5-chloromethylcyclopentadiene gives only 1-methylbenzvalene (33). This shows that the carbene adds to the 1- and 4-positions of the diene component. If the addition occurs at the 1- and 2-positions, the 2-methyl isomer should be formed. The reaction of lithium methylcyclopentadienide with methylene chloride and methyllithium affords 2- and 3-methylbenzvalene both of which are formed by the attack of methylene chloride or chlorocarbene at the unsubstituted carbon of the cyclopentadienide (34). 5-Methyl-5-cyclopentadienylcarbene was shown by MINDO/3 calculation to have a bisected conformation which seems to be susceptible to 1,4-addition [35].

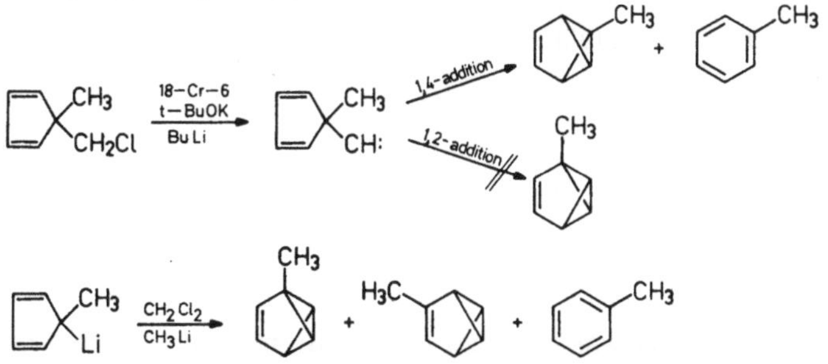

(33)
(34)

Katz's procedure may widely be applied; some examples are given in Burger's review [36]. Starting from dicyclopenta(a,d)benzene dianion, a bisbenzvalenobenzene was prepared as described by scheme (35) [37]. The bisbenzvalene is isomerized in different steps to anthracene by catalysis with silver ions while the benzvalene part is converted to a methylenecyclopentadiene by thermolysis.

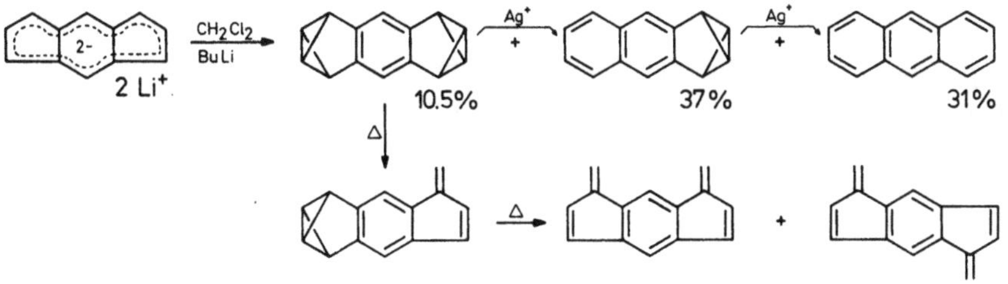

(35)

Among perfluorinated aromatic compounds, only hexakis(trifluoromethyl)benzene was converted to benzvalene. Using an ultraviolet lamp of longer wavelength (307 nm), benzvalene was selectively obtained in high yield (12) [14,15]. This benzvalene is very stable and relatively easy to handle. Its reactions will be discussed in Section 2.5.

2.3 Syntheses of Prismanes

Most prismanes were synthesized by photoreaction of benzenes. The formed intermediates were supposed to be Dewar benzenes which were cyclized by a (2+2)addition reaction. An early example is the photolysis of tri-*tert*-butylbenzene as mentioned before. In this case, a mixture of the prismanes was isolated.

The photolysis of hexamethyl Dewar benzene gives hexamethylprismane (36) [38].

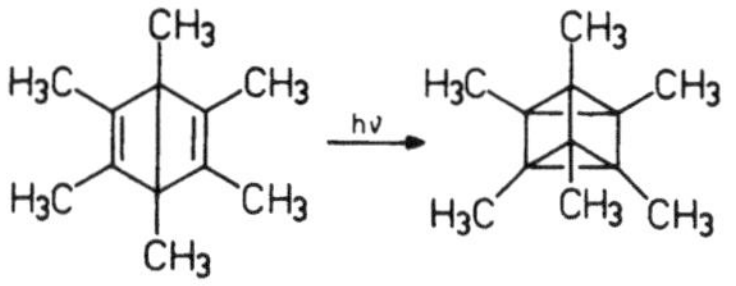

(36)

Unsubstituted prismane was synthesized by Katz's group [39]. Benzvalene was reacted with 1-phenyl-1,2,4-triazolinedione to yield an adduct whereby the benzvalene behaves like a diene. The adduct was hydrolyzed, oxidized and denitrogenated to prismane (37). In spite of its high strain, this prismane is fairly stable in solution ($t_{1/2}$ = 11 h at 90° in d-toluene) whereas a pure sample explodes violently when scratched with a spatula.

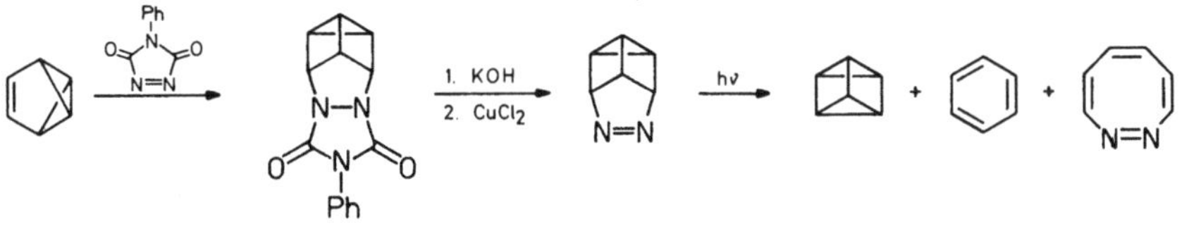

(37)

Perfluoroalkylated Dewar benzenes may easily be converted into their prismane counterparts by irradiation with short wavelength ultraviolet light (38) [14,15,21].

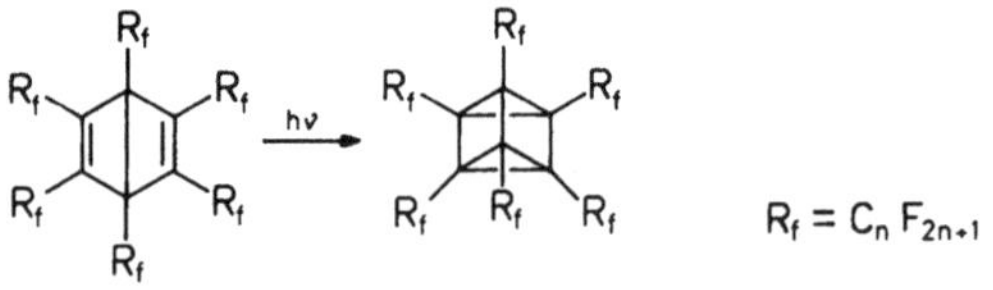

(38)

The photolysis of perfluoro(1,3,5-trimethyl)benzene yields perfluorotrimethylprismane [19]. The latter was considered to be formed via a Dewar benzene. The prismane was isomerized to another Dewar compound which is an intermediate of perfluoro(1,2,4-trimethylbenzene) (see Eq. 16). There are many reactions where prismanes are postulated as intermediates. Some of them are described in the review of Bryce-Smith and Gilbert [29].

2.4 Reactions of Dewar Benzenes

The reactions of Dewar benzenes can be classified into three types:
1. isomerization to benzene counterparts,
2. addition to the strained double bonds, and
3. cycloaddition to the double bonds.

2.4.1 Isomerization to Benzenes

Although a Dewar benzene has a highly strained ring system it is fairly stable, and its isomerization to the benzene counterpart proceeds much more slowly than expected from its ring strain. This stability was explained by the conservation of the orbital symmetry by Woodward and Hoffmann [40]. The central bond of the Dewar benzene is only allowed to open in a conrotatory manner which would give *cis,cis,trans*-cyclohexatriene. This process is impossible due to the high strain of the ring and only disallowed disrotatory opening to the Kekulé benzene can occur. This disallowance is a barrier which is responsible for the stability of the Dewar benzene (39).

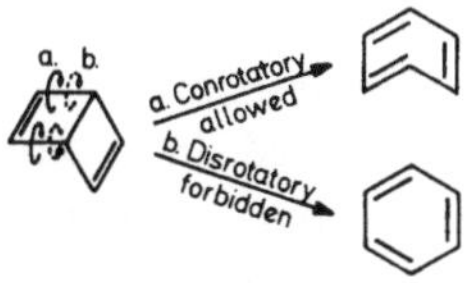

(39)

M. J. S. Dewar calculated the activation energy of the isomerization of an imaginary *trans*-Dewar benzene to benzene and expected that this energy would be much smaller than that of the isomerization of *cis*-Dewar benzene (usual Dewar benzene) since the conrotatory ring opening would give all-*cis*-cyclohexatriene (40, 41) [41]:

There are many theoretical considerations concerning the isomerization of Dewar benzene. For example, Dewar expected chemiluminescence since this process should give electronically excited benzene [42]. This expectation was confirmed by Turro and his coworkers [43].

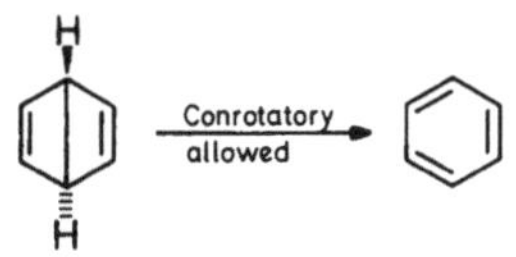

$E_a^{\ddagger}$ = 4.5 kcal/mol (MINDO) (40)

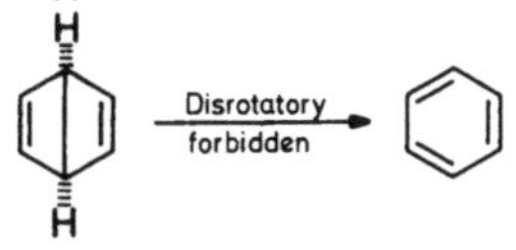

$E_a^{\ddagger}$ = 20.4 kcal/mol (found) (41)

To investigate the structure of the transition state in the isomerization, the activation volumes of the isomerization of Dewar benzene and hexamethyl Dewar benzene (42, 43) were measured [44]:

$\Delta V^{\ddagger} = +(5 \pm 2)\ cm^3/mol$ (42)

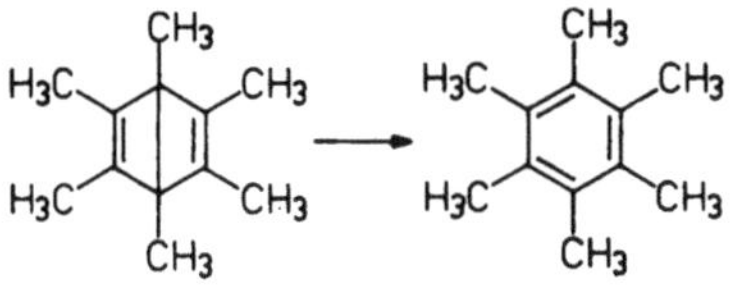

$\Delta V^{\ddagger} = -(12 \pm 3)\ cm^3/mol$ (43)

The low positive value for Dewar benzene is consistent with the mechanism according to which first breaking of the central bond occurs. If the isomerization of the hexamethyl Dewar benzene proceeds by the same mechanism, the negative value is rather strange. This might be due to the fact that the six methyl groups become more crowded in the near-planar transition state than in the starting material.

Isomerization of hexamethyl Dewar benzene to hexamethylbenzene proceeds photochemically, some nitriles being used as sensitizers. In the photoisomerization in the presence of 1-cyanonaphthalene, an emission from exciplex has been observed [45]. Furthermore, fluorescence of nitriles like telephthalonitrile is extinguished by hexamethyl Dewar benzene which isomerizes to the benzene [46]. This fact shows that the singlet excited state of nitriles participates in the formation of the exciplex. The photosensitized isomerization of hexamethyl Dewar benzene in the presence of fumaronitrile is affected by the polarity of the solvent used. In a non-polar solvent the quantum yield is low while it is higher in a polar solvent. This fact suggests that this isomerization follows a radical-ion chain mechanism (44) [47]:

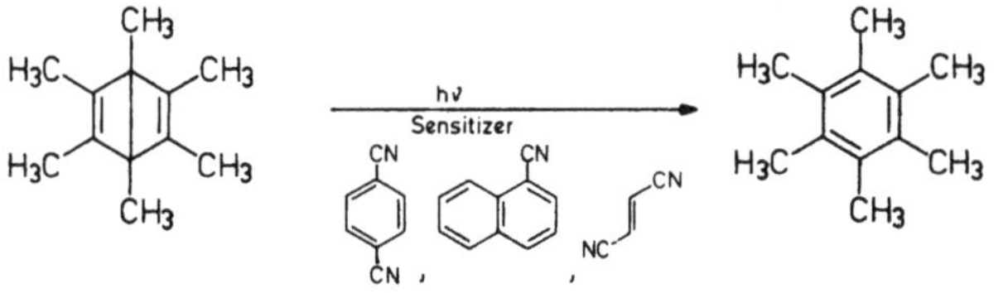

(44)

Non-sensitized isomerization of unsubstituted Dewar benzene and benzvalene was found to proceed in high quantum yield (45, 46) [48].

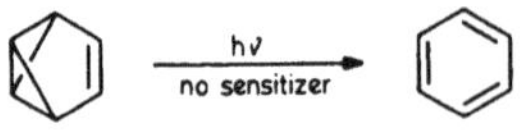

(45)

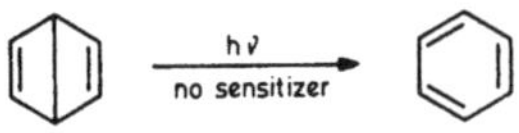

(46)

The treatment of 1-chloro-2-methyl Dewar benzene with lithium diethylamide causes aromatization and introduction of a diethylamino group with elimination of chlorine. The mechanism of this reaction was established on the basis of the results obtained by use of deuteriodiethylamine. Interesting is that a bicyclohexatriene intermediate was proposed [49]. The introduction of deuterium was considered to be due to the 1,4-addition of deuteriodiethylamine to the strained ring. Undeuterated products are formed by direct substitution which is a competitive process (47).

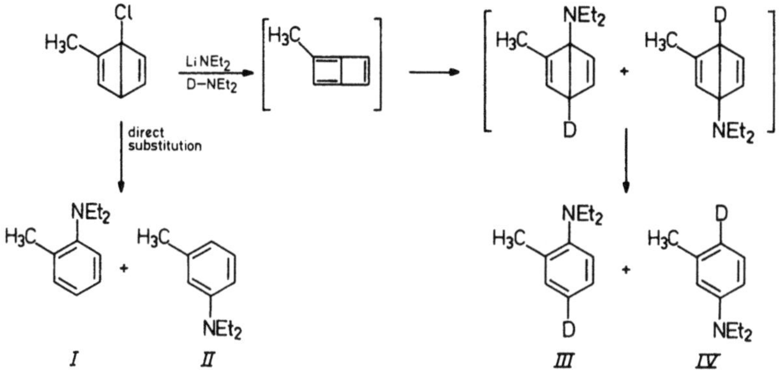

(47)

I + *III* = 76–82% (D% 14%)
II + *IV* = 24–18% (D% 42%)

The Dewar isomers of highly fluorinated benzene derivatives are very stable compared with their hydrocarbon counterparts. The high stability is confirmed by the fact that hexakis(pentafluoroethyl)benzene can be converted to its Dewar isomer by flash thermolysis.

Lemal studied the isomerization of hexakis(trifluoromethyl)Dewar benzene and found that the half-life of the compound is much longer (4 h at 80°) than that of hexamethyl benzene (48 h at room temp.) [50].

The thermal isomerization of tetrafluoro Dewar benzenes was investigated in detail (48) [51].

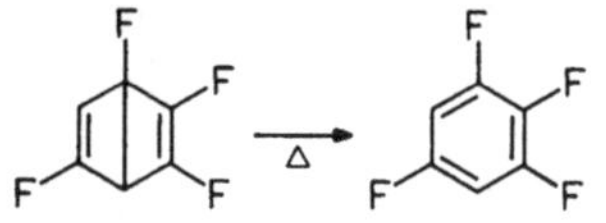

(48)

2.4.2 Addition Reactions

The double bonds of Dewar benzenes are isolated highly strained double bonds. Therefore they undergo usual addition reactions more easily than non-strained olefins.

Non-polar reaction of Dewar benzenes usually afford addition products.

(49) [3]

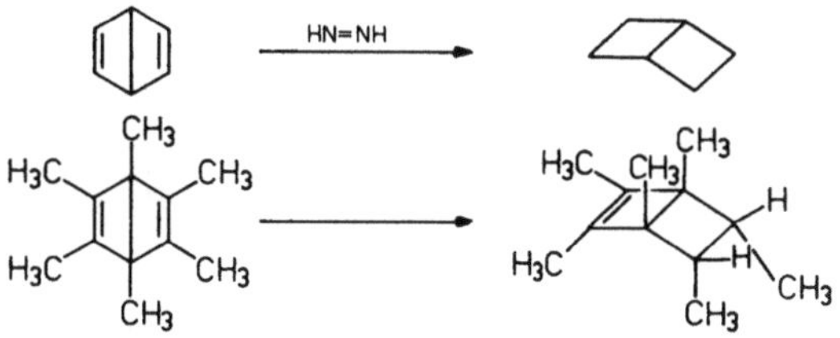

(50) [52]

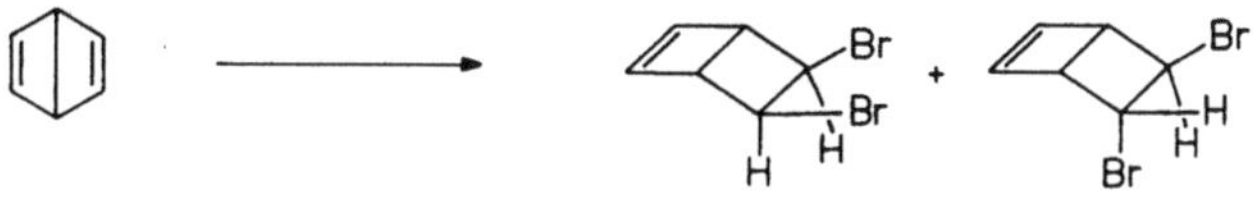

(51) [53]

Characteristic is that Dewar benzene yields a mixture of a *cis*- and *trans*-dibromide (51). This result show that the *endo*-side of Dewar benzene is more crowded than the *exo*-side. Thus, molecular orbital estimations (STO-3G) revealed that the protons on the olefinic double bond of Dewar benzene are shifted into the *endo* position (52)[54].

(52)

In contrast to homolytic additions to Dewar benzenes, polar additions are accompanied by rearrangement of the carbon skeleton:

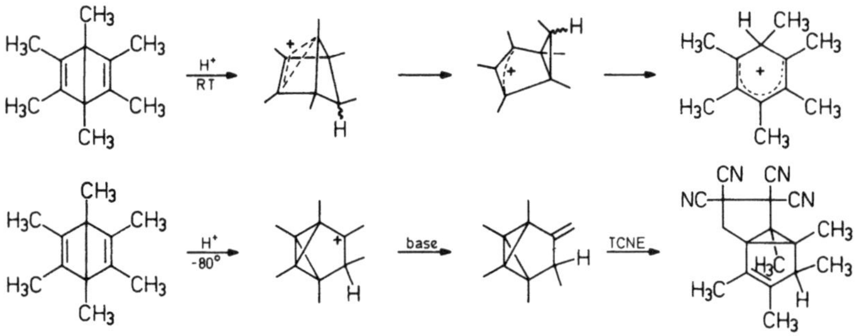

(53)

(54)

Thus, the treatment of hexamethyl Dewar benzene with acid at ambient temperature causes rearrangement to a phenonium ion (55)[55] while at lower temperature a cation is formed which is converted into an *exo*-methylene compound on treatment with a base[56]. Addition of tetracyanoethylene to the latter yields an 1,4-addition product (54).

In the case of a homolytic addition reaction of *exo*-methylenecyclohexadiene, no rearrangement occurs (55)[57].

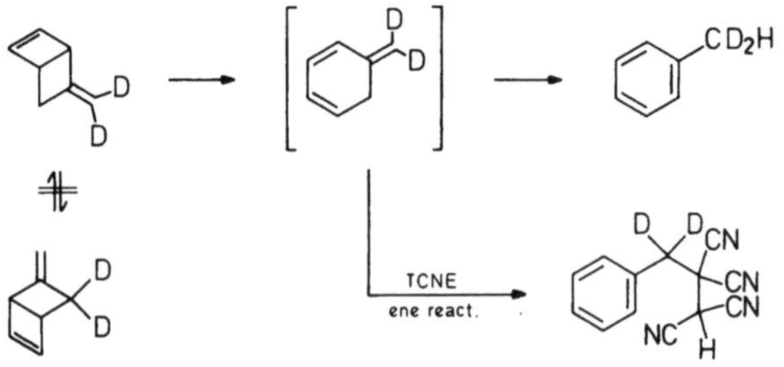

(55)

In contrast to the reaction of 5-exomethylenebicyclo[2.2.0]hexene-1, its pentamethyl analog reacts with tetracyanoethylene to give the Diels-Alder adduct rather than the ene reaction product (56)[58].

The cycloaddition reaction at the double bond of the Dewar benzene occurs much more easily than at a usual olefinic double bond, probably due to the high strain.

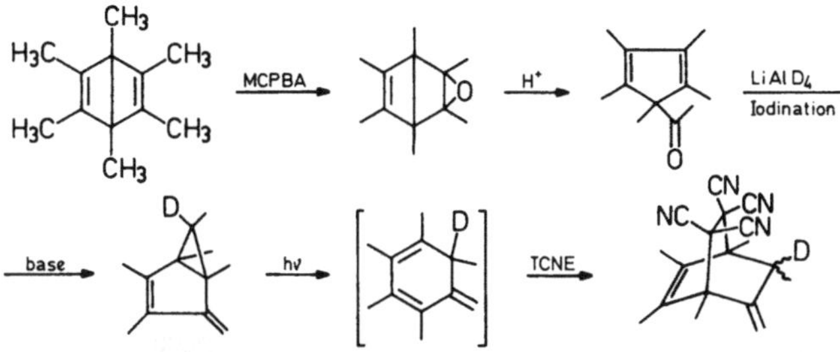

(56)

The 1,3-dipolar addition of ozone to hexamethyl Dewar benzene gives the corresponding ozonide, which is photolytically converted to tetramethylcyclobutadiene (57) [59]. The photochemical addition of benzaldehyde to hexamethyl Dewar benzene yields a (2+2)cycloadduct the oxetan part of which is reduced much more rapidly than the cyclobutenic double bond. This may be due to the high substitution around the double bond (58) [60].

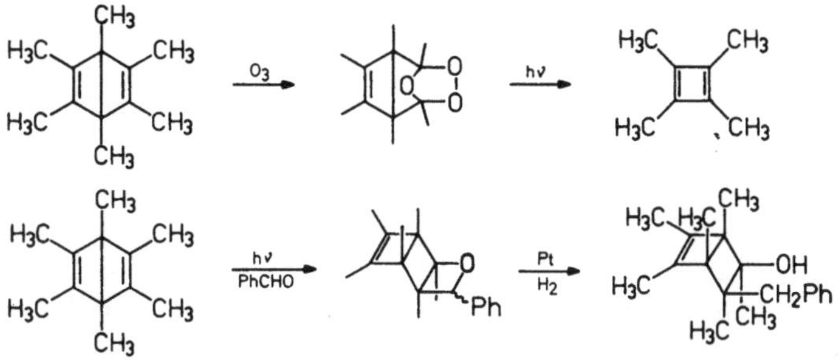

(57)

(58)

Similar addition reactions have been reported for perfluorinated Dewar benzenes, although few electrophilic reactions are known since the high electronegativity of fluorine makes the double bond much less nucleophilic. Thus, bromine adds to the double bonds of hexafluoro Dewar benzene in a stepwise manner (59) [13].

Nucleophiles easily added to the double bond of perfluorinated trimethyl Dewar benzenes followed by elimination of a fluoride ion. A similar reaction was reported

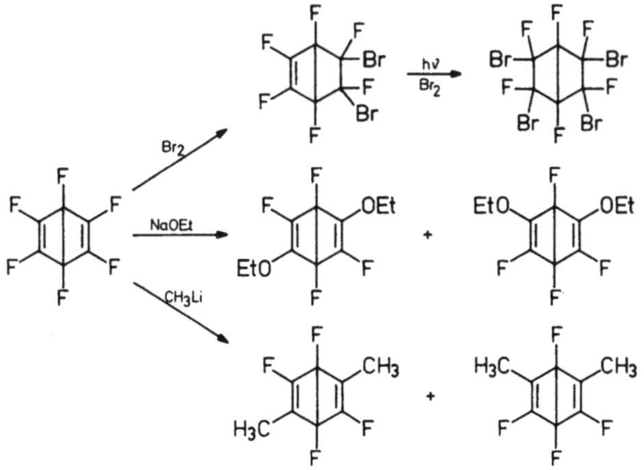

(59)

by Haszeldine's group (60)[19]. In both examples, the fluorine at the bridgehead is not substituted. The high reactivity of the double bond was utilized in the reaction

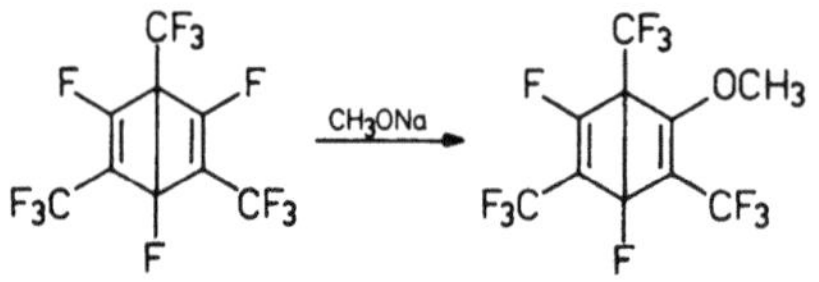

(60)

of hexafluoro Dewar benzene with $CF_2(OF)_2$ for the synthesis of a special polymer (61)[61]. The photochemical addition of the same reagent to hexafluoro Dewar benzene yields monomeric products including perfluorooxepine (62)[62]:

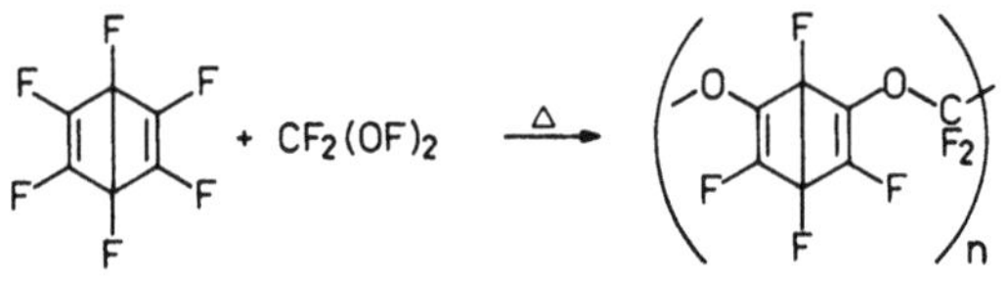

(61)

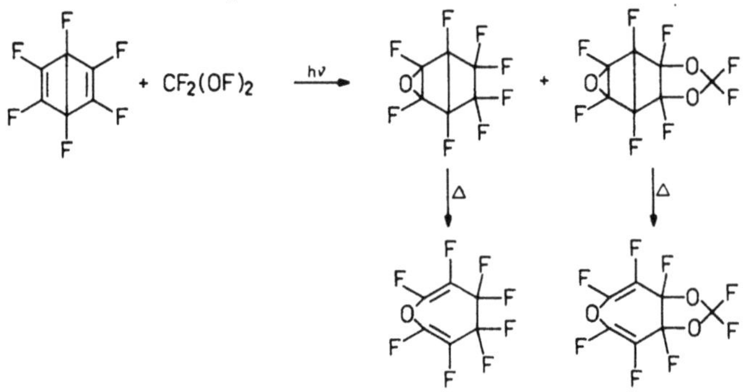

(62)

The double bond of perfluorinated Dewar benzenes is much more reactive as a dienophile in the Diels-Alder reaction than that of the non-fluorinated Dewar benzenes since the former is considerably more electronegative than the double bond of the latter. Haszeldine and his coworkers investigated the Diels-Alder reaction of perfluorinated Dewar 1,3-xylene and established order of reactivity of fluoroalkene part of Dewar benzene shown in Scheme (63). This order results from the electronegativities of the substituents. Furthermore, hexakis(trifluoromethyl)Dewar benzene the double bonds of which are substituted by trifluoromethyl groups was reported not to undergo a Diels-Alder reaction, probably because of the bulkiness of the trifluoromethyl groups. Nevertheless, the electronic effect of the substituents should activate the double bonds. Thus, a reinvestigation of the Diels-Alder reaction revealed that the latter reacts above room temperature with 1,3-dienes. The trifluoromethyl groups stabilize the strained ring system sufficiently so that it does not isomerize to the Kekulé isomer even at high temperatures (64) [64].

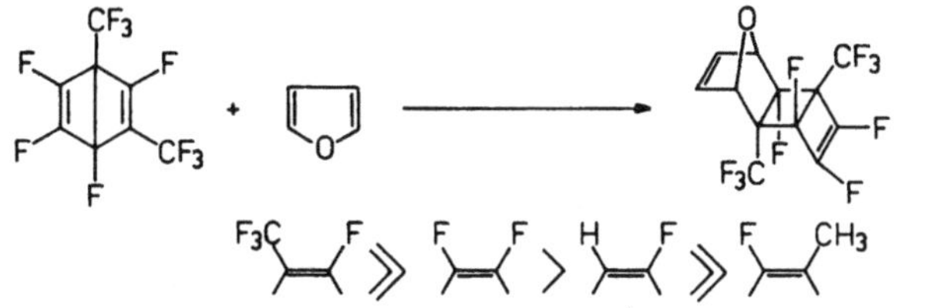

(63)

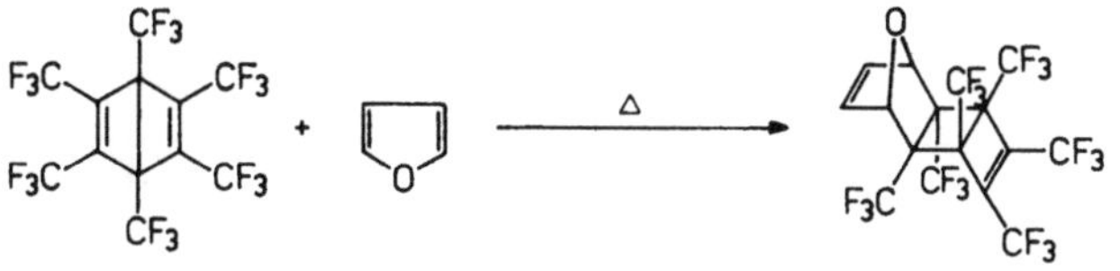

(64)

The Diels-Alder reaction of perfluorinated Dewar benzenes has been applied to the synthesis of many types of compounds. For a example, a fluorinated naphthalene was synthesized by this method (65) [65]. The Dewar benzene can therefore be considered as an activated form of benzene like a benzyne.

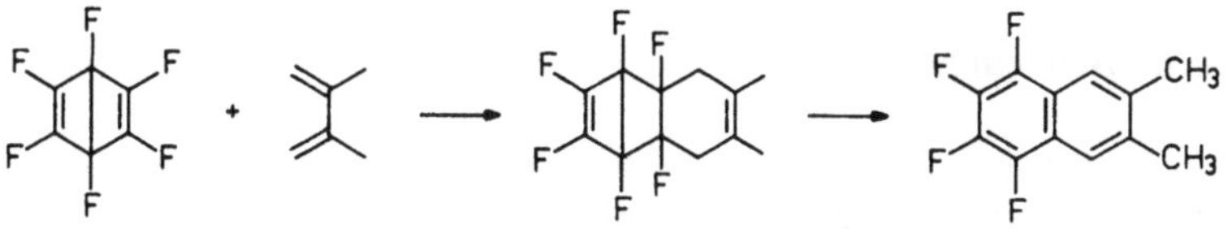

(65)

Also metal complexes of hexafluoro Dewar benzene may be subjected to a Diels-Alder reaction, the central bond remaining unaffected (66) [66]:

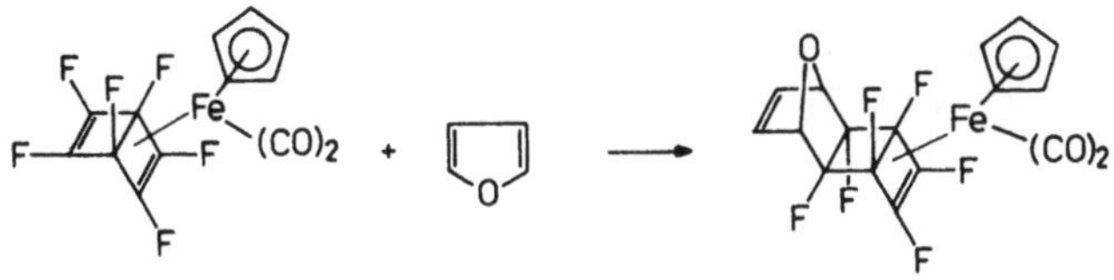

(66)

These fluorinated Dewar benzenes are also reactive as dipolarophiles in the 1,3-dipolar reactions [63].

The photolysis of hexafluorobenzene in the presence of oxygen yields the corresponding Dewar benzene and its epoxide; the latter may act as a dienophile and thermally rearrange to a perfluoro-1,3-cyclohexadienone (67) [66]. This reaction is in marked contrast to the reaction of hexamethyl Dewar benzene (see Eq. 56).

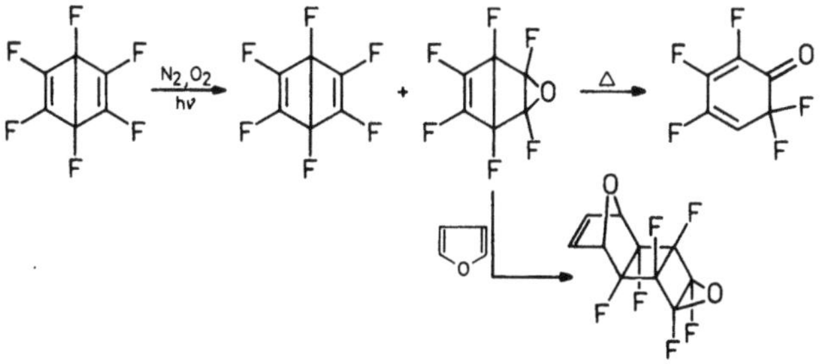

(67)

The photolysis of tetrafluorocyclobutenedicarboxylic anhydride obtained from hexafluoro Dewar benzene yields tetrafluorocyclobutadiene which is trapped as an adduct with furan (68) [67]. This is another useful method for cyclobutadiene.

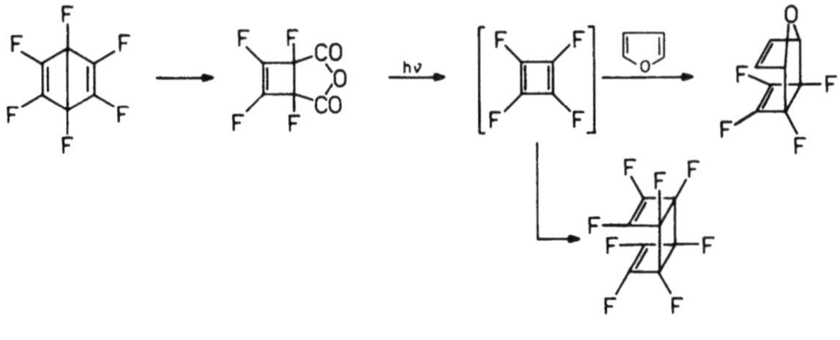

(68)

As mentioned above, Dewar benzenes are valuable intermediates for the synthesis of many kinds of compounds.

2.5 Reactions of Benzvalenes

The reactions of benzvalenes are classified into three categories: isomerization, reactions at the double bond and those of the bicyclobutane part.

According to Burger, the isomerizations are grouped into the following three types (69, 70, 71) [36]:

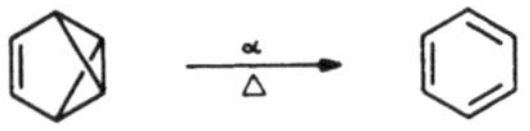

(69)

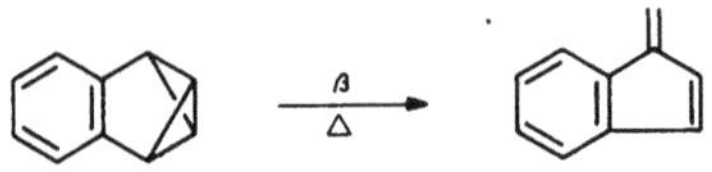

(70)

(71)

An α-type reaction is the isomerization to the Kekulé benzene which is accomplished thermally or catalyzed by silver ions (72) [43,68].

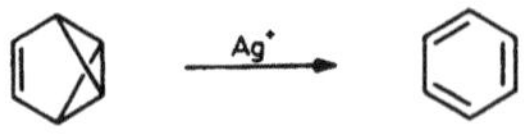

(72)

The study of the mechanism using 1-deuteriobenzvalene revealed that the first attack of the silver ion occurs reversibly since the redistribution of deuterium takes place within the half-life period of benzvalene [69]. The numbers under the formulas indicate the ratio of the deuterium in the three deuterated benzvalene isomers (73).

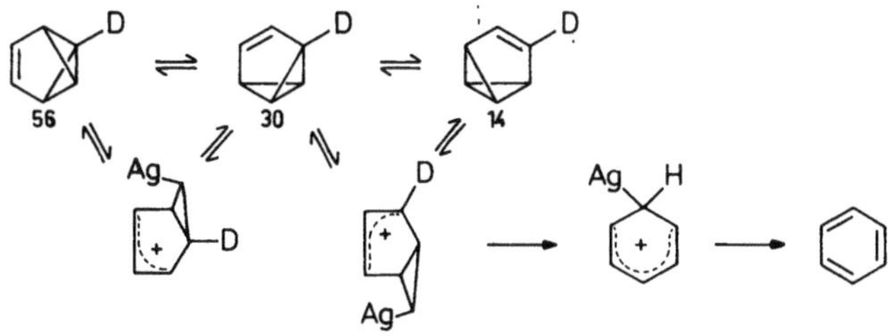

(73)

A similar result was obtained by the reaction of 1,6-dideuteriobenzvalene (74) [70]. The isomerization of deuterated benzobenzvalene catalyzed by silver ions is not accompanied by rearrangements (75) [69].

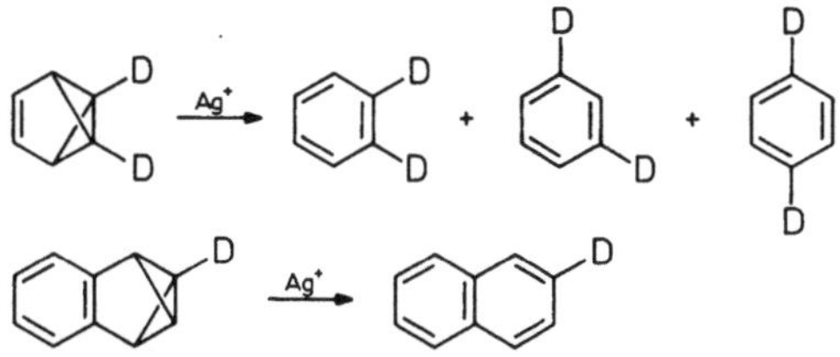

(74)

(75)

It can be efficiently performed by photolysis if oxygen is strictly excluded as mentioned before [48].

A β-type rearrangement is catalyzed by metals such as copper, platinum, palladium, or gold. The copper-catalyzed rearrangement of 1-deuteriobenzvalene has been extensively investigated. This rearrangement is supposed to proceed via a cyclopentadienylcarbene complex which undergoes a rapid 1,5-shift of the carbene moiety. 60% of the deuterium have been found in the methylene part and 40% are distributed in equal portions among the α- and β-position of the cyclopentadiene part (76) [68].

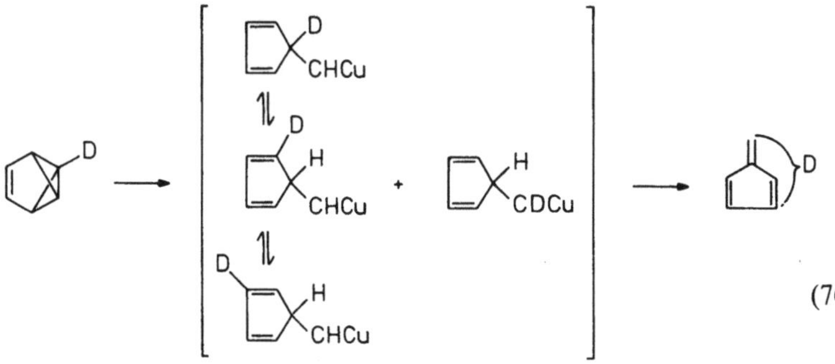

(76)

A γ-type rearrangement of the usual benzvalenes has rarely been observed because a Dewar compound is formed as a rearrangement product which is unstable and isomerizes to a Kekulé compound. In the case of bicyclobutenonaphthalene, two isomerization steps to the corresponding cyclobuteno isomers are allowed but isomerization of the latter to plaiadirene is forbidden due to the conservation of orbital symmetry. Moreover, plaiadirene is no longer stabilized by aromaticity (77) [71].

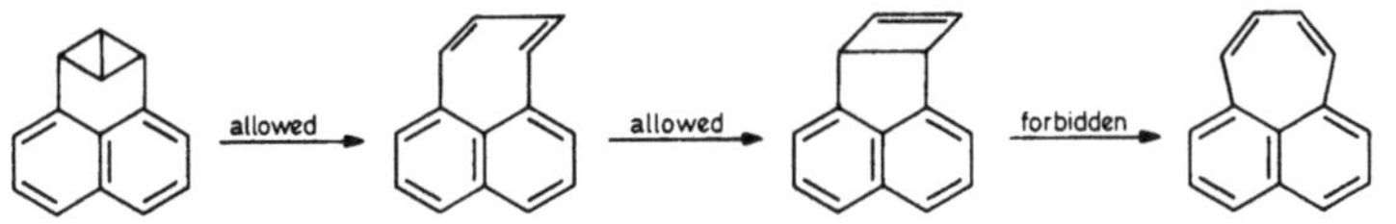

(77)

The reaction of benzvalene with iron pentacarbonyl is supposed to involve attack of iron at the double bond which is different from the silver- or copper-catalyzed reaction (78) [72].

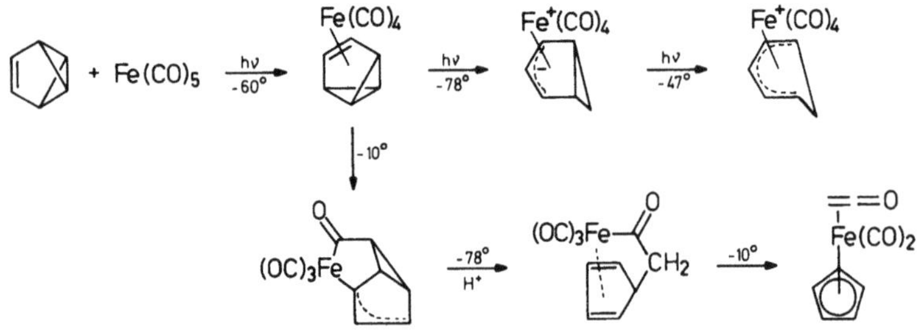

(78)

Hexakis(trifluoromethyl)benzvalene is the only perfluorinated benzvalene which has been known so far. It is very stable at room temperature. The thermal and photochemical α-isomerization of this compound has been reported.

Next, the reaction of the double bond of benzvalene will be discussed. Benzvalene itself may be reduced with diimine whereby the bicyclobutane part remains unaffected (79) [73].

(79)

The double bond of unsubstituted benzvalene is considered to be an electron-rich system which reacts with electron-deficient 1,3-dienes; namely the Diels-Alder reaction is reverse electron demanding. Two Diels-Alder reactions of benzvalene are described in the following. Benzvalene does not react with usual 1,3-diene but with electron-deficient 1,3-dienes like 1,2,4,5-tetrazine (80) or hexachlorocyclopentadiene (81) [74]. Dihalogenocarbene also adds to the double bond (82) [75]. The obtained

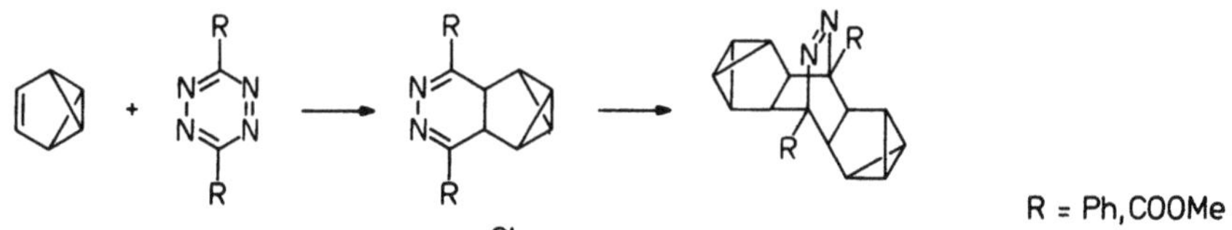

(80)

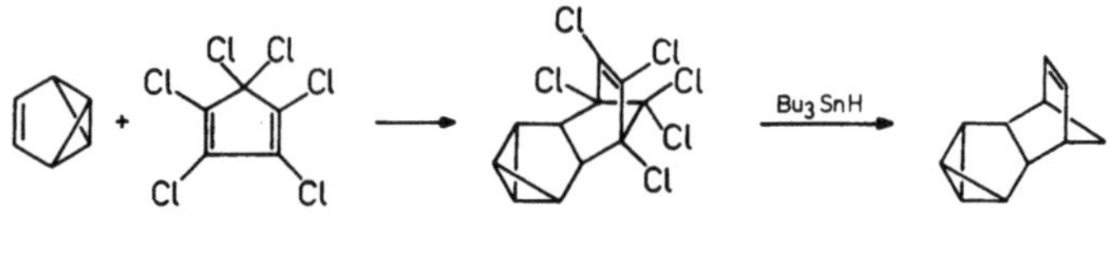

(81)

gem-dihalo compounds are reduced with Bu_3SnH. The 1,3-dipolar reaction of benzvalene with azides gives 1,2,3-triazolines, the thermolysis of which yields different types of compounds, depending on the substituent on the nitrogen atom (83) [78]. Diazo compounds may also add to the olefinic double bond (84) [80]. With dichloroketene benzvalene undergoes a (2+2)cycloaddition reaction. The resulting adduct is converted into a cyclobutene compound which is isomerized to cyclooctatetraene [81] or reacted with 1,2,4-triazolinedione to form an interesting cage compound (85) [82].

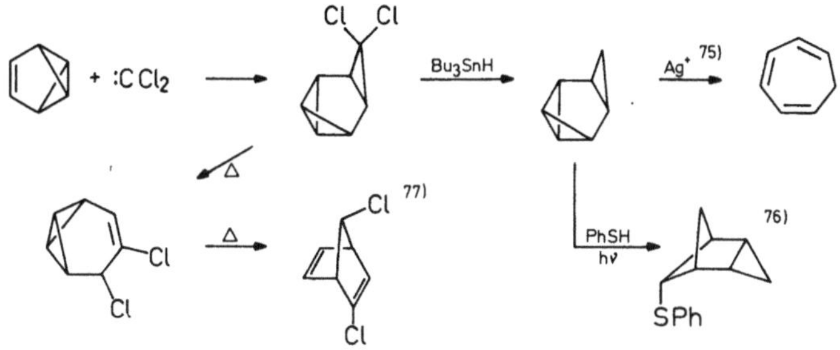

(82)

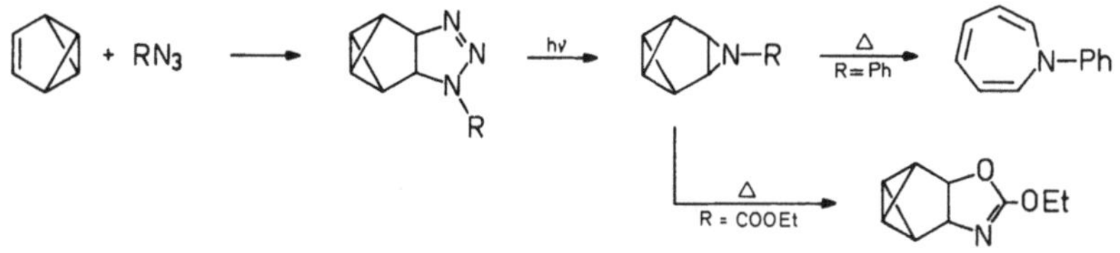

(83)

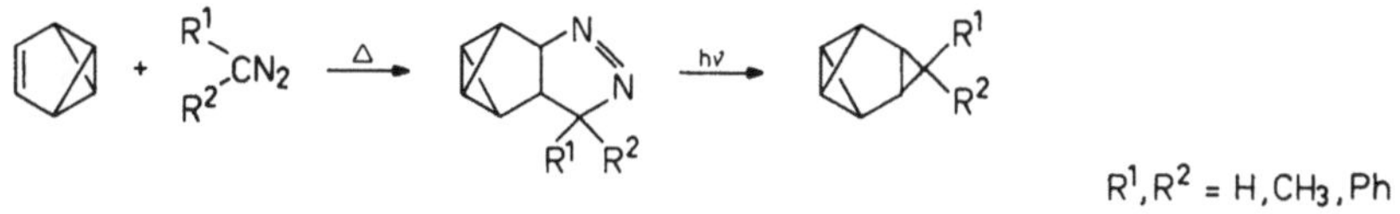

(84)

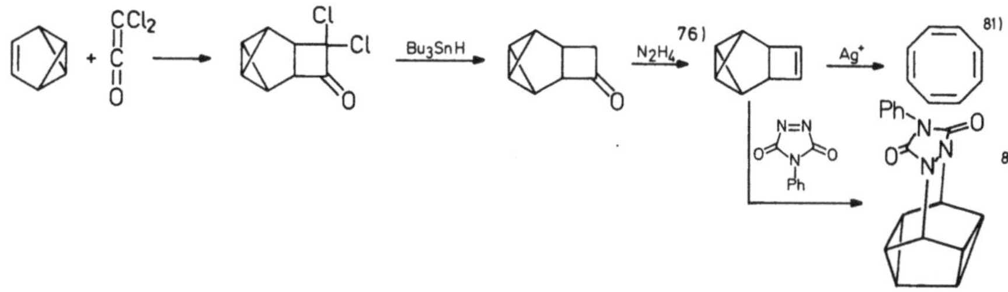

(85)

Ozone and singlet oxygen add to the benzvalene double bond to form the corresponding ozonide and 1,2-dioxetane adduct respectively with retention of the bicyclobutane part (86) [79].

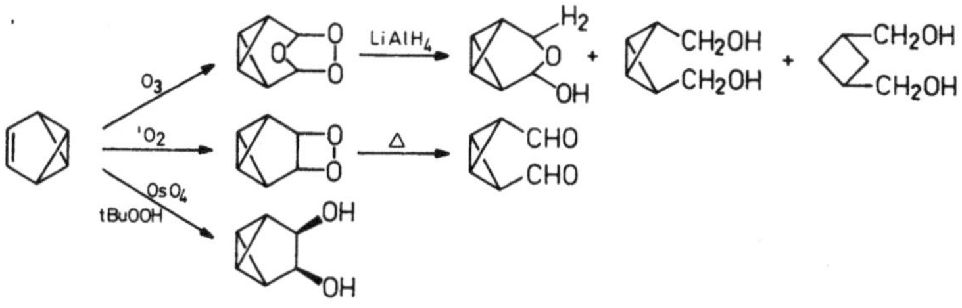

(86)

0,0-Dimethyl dithiooxalate reacts as a diene with benzvalene to form a 1,4-dithiine cycloadduct (87) [83].

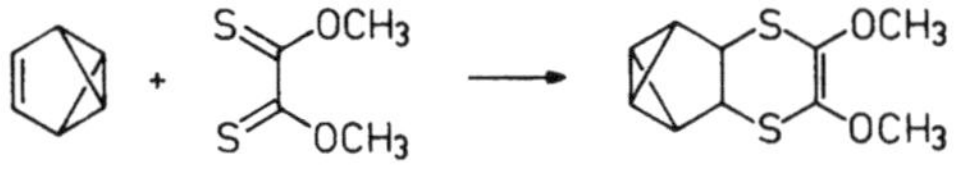

(87)

The reaction of 1-phenyl-1,2,4-triazolinedione occurs at the 1,4-position of the vinylcyclopropane system of benzvalene. This reaction has been used for the synthesis of prismane as mentioned before [39] (see Eq. 37).

The addition of bromine to benzvalene gives a dibromide whereby the attack of the halogen seems to occur at the bicyclobutane part, but the experiment using isotopes showed that bromination starts on the double bond followed by a rearrangement (88) [84].

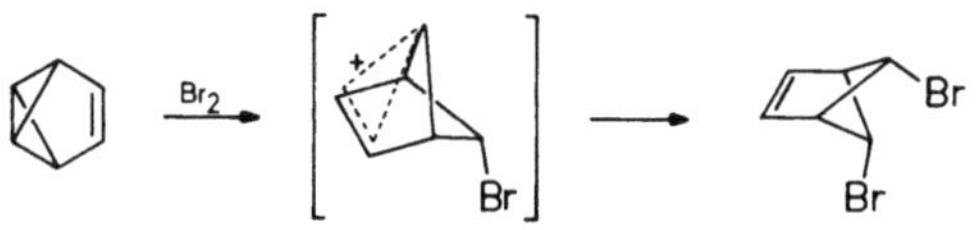

(88)

The addition of chlorine also yields a "bicyclobutane-opened" dichloride (89) [85] while iodination occurs either at the bicyclobutane part or double bond depending on the experimental conditions (90, 91) [86].

(80)

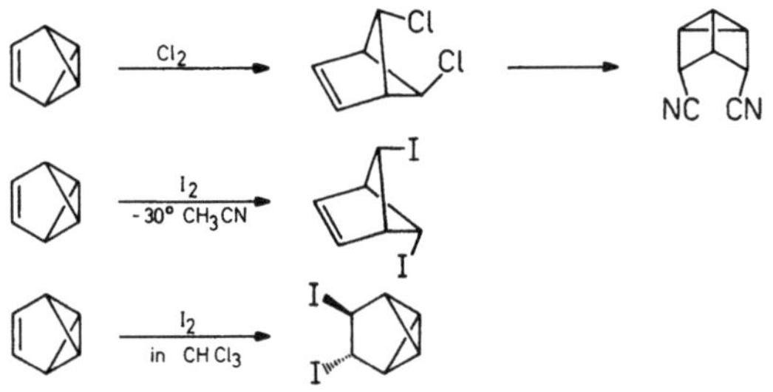

(90)

(91)

Katz and his coworker showed that phenylsulfenyl chloride adds to the double bond without any isomerization. This is explained by the high stability of the generated episulfonium ion (92) [84].

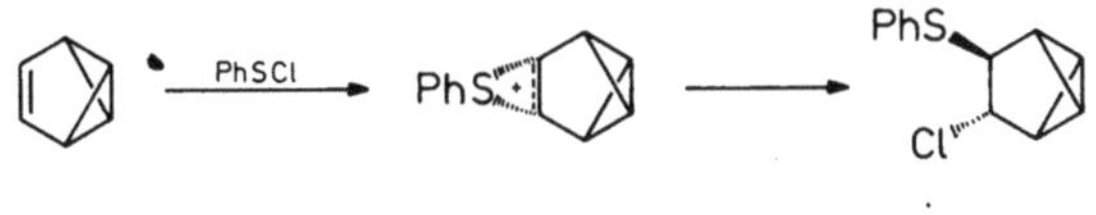

(92)

The reactivities of the double bond and bicyclobutane part with respect to thiophenol depend on the respective ring system. Thus, the double bond of benzvalene is more reactive than the bicyclobutane part while in the case of tricyclo-[4.1.0.0^{5,7}]heptene, the reverse order of reactivity is observed (93, 94) [88].

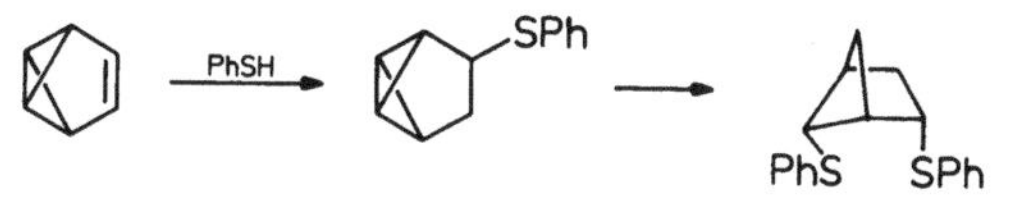

(93)

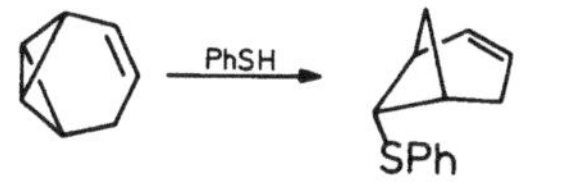

(94)

Some interesting analogs of methylated benzvalene were obtained from hexamethyl Dewar benzene. Although their ring system does not contain benzvalene, some of the most interesting reactions are briefly described (95).

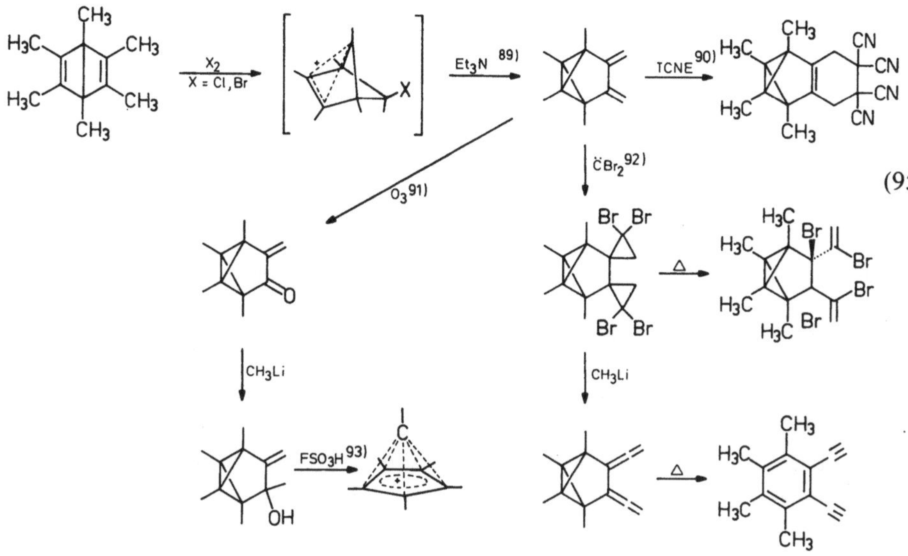

(95)

Next, some reactions of hexakis(trifluoromethyl)benzvalene will be discussed. This compound is very stable as mentioned before and its olefinic double bond reacts as an efficient dienophile. The difference in the reactivity between unsubstituted benzvalene and the hexakistrifluoromethylated derivative is that the former is nucleophilic while the latter is highly electrophilic. Thus, the former does not react with usual 1,3-dienes, in contrast to the latter. Here, mainly describe our results of the several reactions of hexakis(trifluoromethyl)benzvalene which were designed for the synthesis of tetrakis(trifluoromethyl)tetrahedrane.

Hexakis(trifluoromethyl)benzvalene reacts as an efficient dipolarophile with phenyl azide, diazomethane and ozone to give the corresponding cycloadducts (96).

The attempted elimination of five-membered heterocycles from the adducts formed from phenyl azide or diazomethane and hexakis(trifluoromethyl)benzvalene was unsuccessful. The photolysis of the ozonide of the benzvalene at room temperature gives the dimer of tetrakis(trifluoromethyl)cyclobutadiene. When the irradiation is carried out in a matrix at −196 °C, the reaction mixture was coloured yellow. The

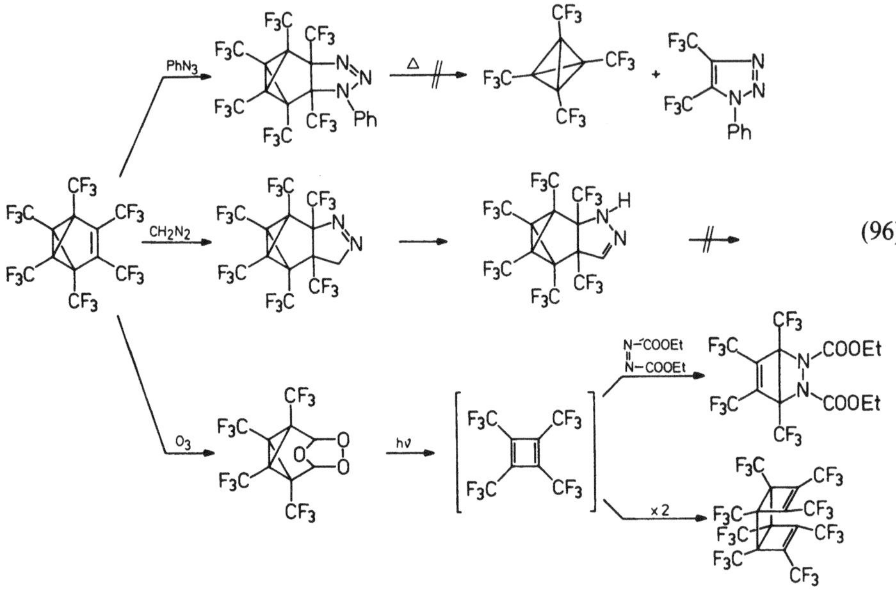

(96)

addition of diethyl azodicarboxylate gives the corresponding cyclobutadiene adduct on warming up [94].

In 1976, Chapman repeated our experiment at 8K and reported that tetrakis-(trifluoromethyl)tetrahedrane is formed. Some time later, Masamune and his co-workers detected cyclobutadiene in the IR and UV spectrum but not the tetrahedrane [95].

The Diels-Alder reaction of hexakis(trifluoromethyl)benzvalene with furan had been reported [63] but not its stereochemistry. We have examined this reaction and found that it proceeds smoothly with many 1.3-dienes if the steric requirement of the olefin is not large in the transition state where *exo*-attack occurs as confirmed by X-ray analysis (97) [96]. Therefore, 2,3-dimethylbutadiene reacts smoothly with hexakis(trifluoromethyl)benzvalene whereas 1,3-pentadiene reacts very slowly and 2,4-hexadiene does not at all. The cycloaddition to cyclobutadiene gives a mixture of *endo*- and *exo*-adducts (98) [97]:

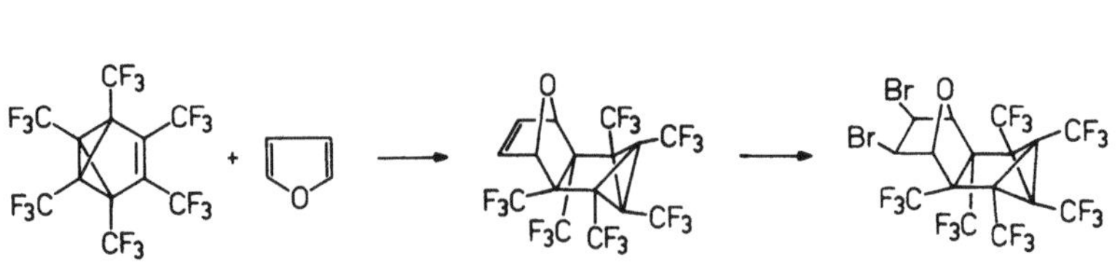

(97)

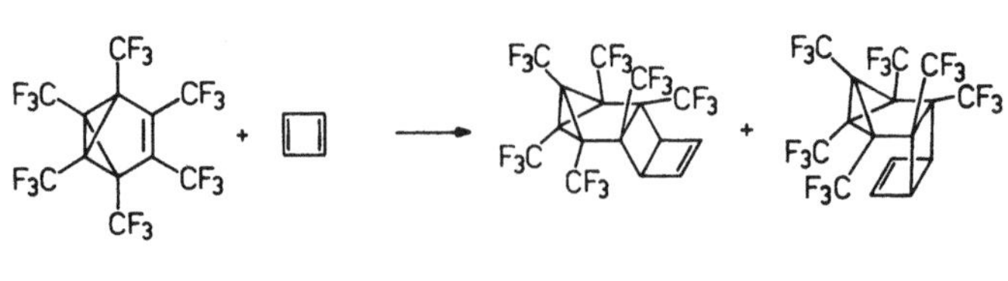

(98)

The former assignment of the structures of the cyclobutadiene adducts of hexakis-(trifluoromethyl)benzvalene by Warrener was corrected by the X ray analysis of the bromination product of the *exo*-adduct [98]. The reaction of the adduct, formed from hexakis(trifluoromethyl)benzvalene and furan, with bromine yields a *cis*-adduct, probably because of the high steric requirement of the trifluoromethyl groups on the bridge-head.

The adduct generated from the benzvalene and butadiene was brominated and dehydrobrominated to a cyclohexadiene which seems to be a suitable precursor of tetrahedrane. However, its photolysis yields (2+2)intramolecular cycloadducts and its thermolysis gives only a cyclobutadiene dimer (99) [99].

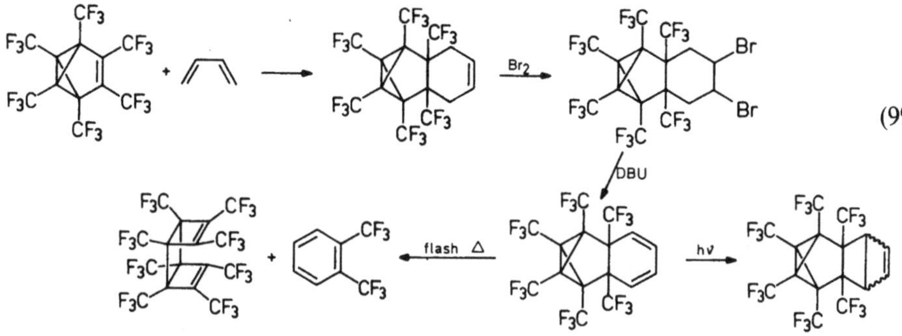

(99)

The ozonide of hexakis(trifluoromethyl)benzvalene is reduced with triphenylphosphine to a dicarbonyl compound. The attempted conversion of this diketone to the tetrahedrane either directly or via halogeno compounds was unsuccessful, but further treatment of the diketone with triphenylphosphine yields new valence-bond isomers of oxepin (100) [100]. The reactions of these isomers will be described in Section 3.4.

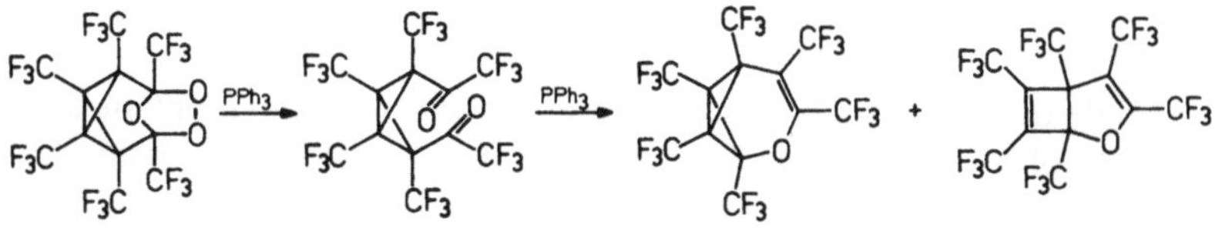

(100)

The photoreaction of hexakis(trifluoromethyl)benzvalene with acetylenes gives (2+2)cycloadducts. This reaction does not proceed with electrondeficient acetylenes, such as dialkyl acetylenedicarboxylate. This fact shows that interactions between the electron-rich acetylene and the electron-deficient double-bond of the benzvalene play an important role (101) [101].

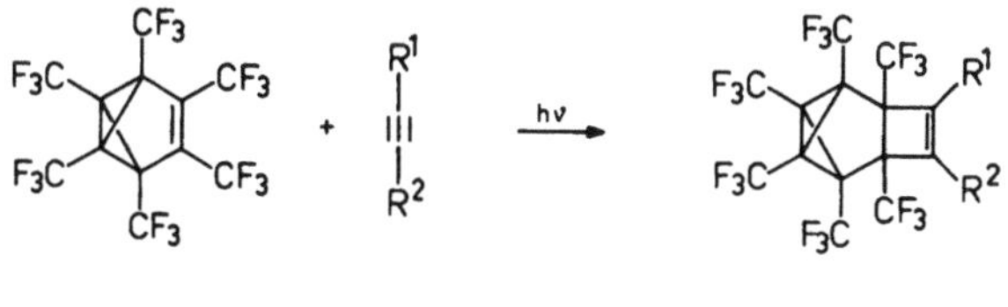

(101)

2.6 Reactions of Prismanes

Prismane is the most highly strained valence-bond isomer of a benzene. Therefore, it isomerizes to the benzene isomer possibly via its Dewar isomer. However, since the activation energy of the isomerization of the Dewar isomer to the benzene is small, the Dewar isomer is rarely observed. Many photochemical isomerizations of benzene derivatives are assumed to proceed via Dewar benzenes and prismanes but in most of cases, such isomers have not been observed.

Other reactions of prismanes are rarely known.

3 Valence-Bond Isomers of Heteroaromatic Compounds

Since a review on Dewar heterocyclic compounds has been published recently [102], in this chapter only some typical reactions of valence-bond isomers of heteroaromatic compounds will be discussed. The emphasis will be placed on the chemistry of trifluoromethylated Dewar thiophene and pyrroles, both of which are rare examples of stable valence-bond isomers of heterocyclic compounds. Since the isolation of valence-bond isomers of heteroaromatic compounds is mostly limited to the perfluoroalkylated isomers, those postulated as intermediates are also included.

3.1 Five-Membered Heterocyclic Compounds

Photoisomerization of five-membered heteroaromatic compounds has been known for a long time and several intermediates have been proposed. An example is the photoisomerization of a 2-substituted thiophene (102).

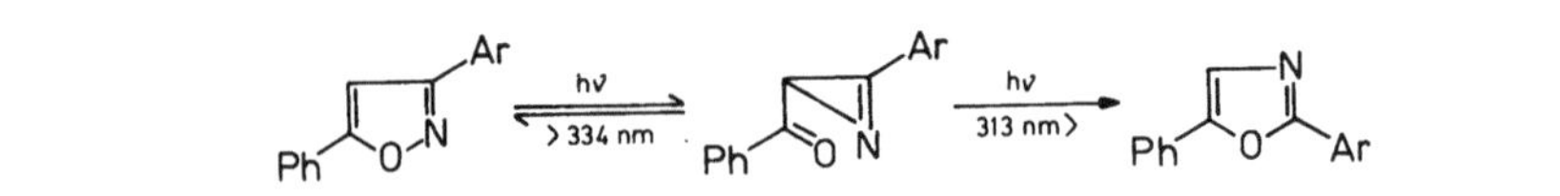

(102)

An early example of the isolation of a cyclopropenyl ketone analog, obtained by photoisomerization of a 3-aryl-5-phenyl-1,2-oxazole, was reported by Ullman and his coworker (103) [103].

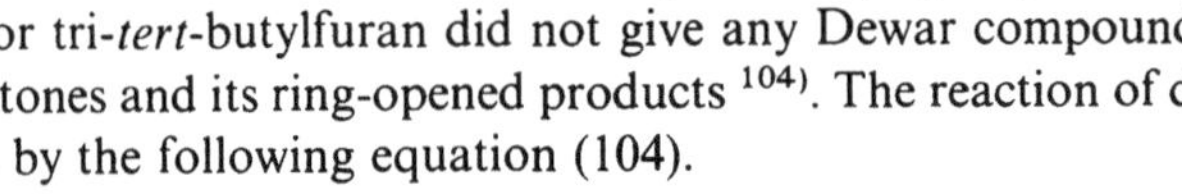

(103)

After the isolation of a Dewar benzene substituted by *tert*-butyl groups, van Tamelen tried to isolate a Dewar furan stabilized by *tert*-butyl groups. However, the photolysis of di- or tri-*tert*-butylfuran did not give any Dewar compound but only cyclopropenyl ketones and its ring-opened products [104]. The reaction of di-*tert*-butylfuran is described by the following equation (104).

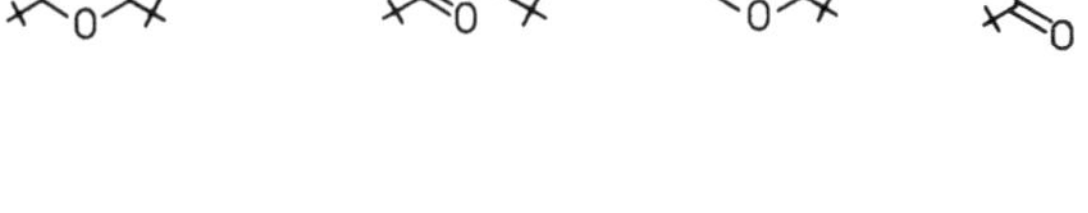

(104)

Another attempt to synthesize a Dewar furan by the photolysis of tetrakis(trifluoromethyl)furan was made since the photolysis of tetrakis(trifluoromethyl)thiophene gives a stable Dewar compound (see Eq. 109). However, even this furan only yields the corresponding cyclopropenyl ketone [105]:

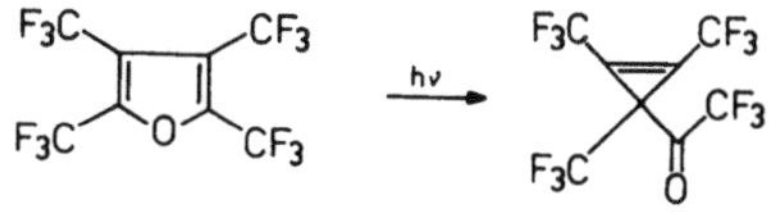
(105)

The photolysis of 2-methylfuran in the presence of a primary amine gives a 1,3-disubstituted pyrrole. This photolysis is assumed to proceed via a cyclopropenyl ketone, which reacts with the amine to form an imine. This imine rearranges to the pyrrole derivative (106) [106]. Couture and his coworker also found that the photo-

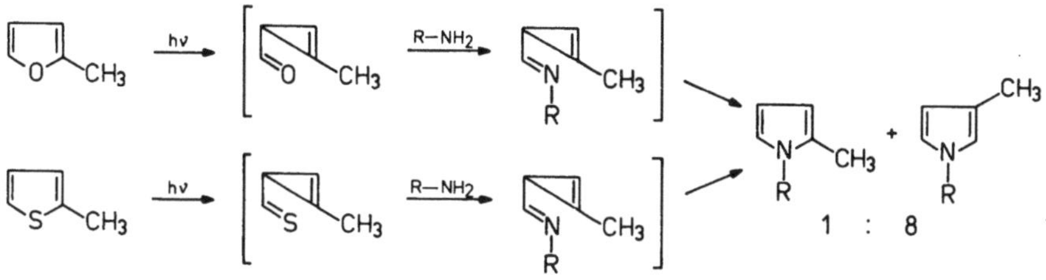
(106)

reaction of 2-methylthiophene in the presence of the same amine gives the same products in the same ratio as in the case of the 2-methylfuran. Therefore, they proposed that the same intermediate imine is formed in the both reactions.

Before this report, Wynberg's group proposed that in the case of 2-substituted thiophenes, the valence shell could be expanded leading to the formation of a dipolar tricyclic intermediate (107) [107].

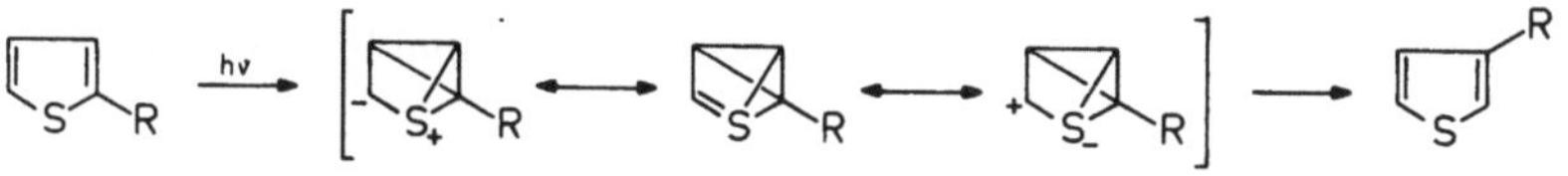
(107)

Later, Kellog, one of the co-authors of Wynberg, postulated a mechanism according to which a two-atomic part of the starting molecule is rotated 90° out of the plane of the remaining atoms (108) [108].

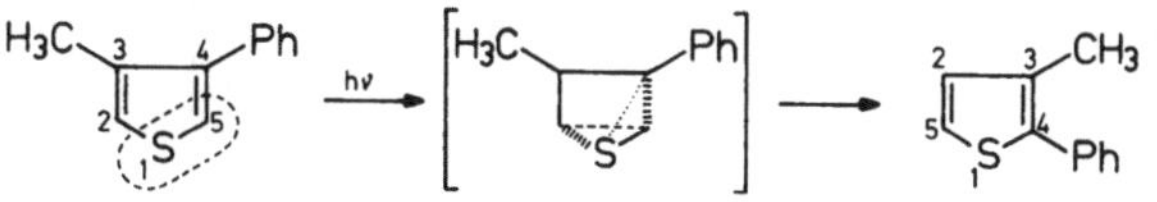
(108)

These intermediates of the photoreaction of thiophenes were not observed directly. The photoreaction of tetrakis(trifluoromethyl)thiophene gives a very stable Dewar isomer [109]. Its structure was elucidated by ^{13}C-NMR and X-ray analysis of the Diels-Alder adduct with tetramethylfuran (109) [110].

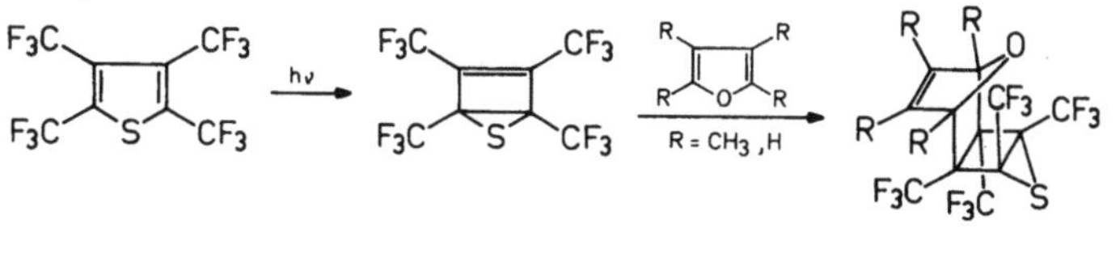
(109)

This is the first example of the isolation of a Dewar isomer of a five-membered heteroaromatic compound. Many interesting reactions of this Dewar thiophene have been reported. Some of them will be discussed in Section 3.2.

The stabilizing effect of the trifluoromethyl groups on the highly strained Dewar compound is still not clear. The Dewar isomer is only formed if the thiophene is substituted by trifluoromethyl groups or other stabilizing groups. The photoreaction of 2-cyanothiophene affords 3-cyanothiophene. If this reaction is carried out in the presence of furan, the adduct of 3-cyano Dewar thiophene and furan is formed. This result suggests that the intermediate of the photoisomerization of 2-cyanothiophene is a Dewar compound (110) [111].

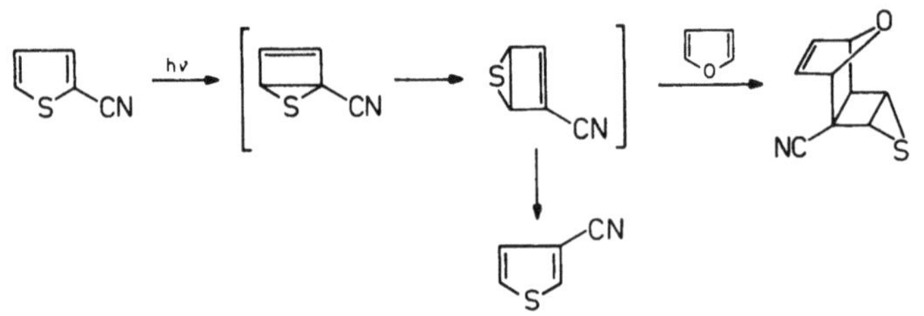

(110)

Furthermore, the photoreaction of 2-cyano-3-methyl- and 3-cyano-4-methylthiophene gives the same Dewar isomer, 3-cyano-4-methyl Dewar thiophene, which can be isolated at room temperature. The Diels-Alder reaction with furan affords a mixture of two adducts. This implies that an equilibrium exists between the three Dewar thiophenes through the migration or walk of the sulfur. The equilibrium is shifted to the 3-cyano-4-methyl Dewar thiophene due to the electronic pull-push effects of the two substituents, although this isomer is less reactive than 3-cyano-2-methyl Dewar thiophene in the Diels-Alder reaction [112].

These results reveal that trifluoromethyl groups are not essential for the formation

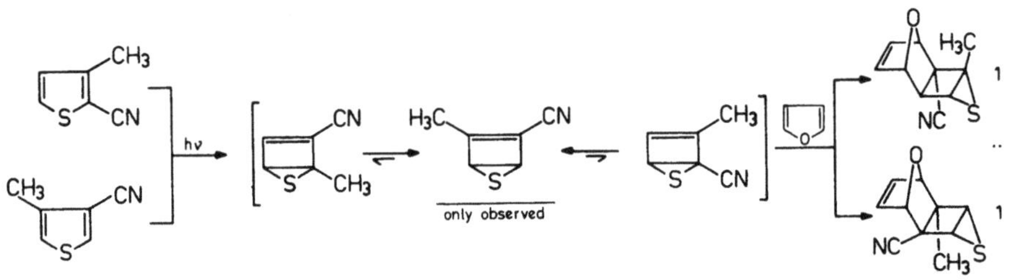

(111)

of the Dewar thiophene. A cyano group is, for example, as electronegative as a trifluoromethyl group. The electronegativity may be an important factor in the formation of the Dewar thiophene. Recently, the present authors have found that 2,3-bis-(trifluoromethyl)thiophene gives a Dewar isomer which reacts with furan to yield two Diels-Alder adducts. This result shows that two trifluoromethyl groups can stabilize a Dewar compound, and that equilibrium exist between the two Dewar isomers (112) [113].

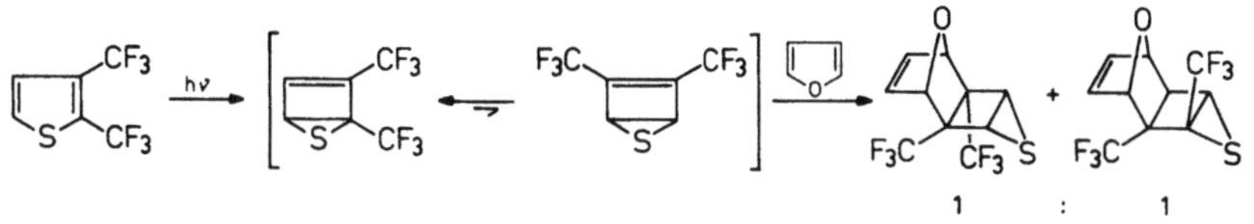

(112)

The cyano group also plays an important role in the photoreaction of 2-cyano-1-methylpyrrole. Hiraoka reported that the photolysis in methanol yields a methanol adduct of a Dewar isomer (113) [114].

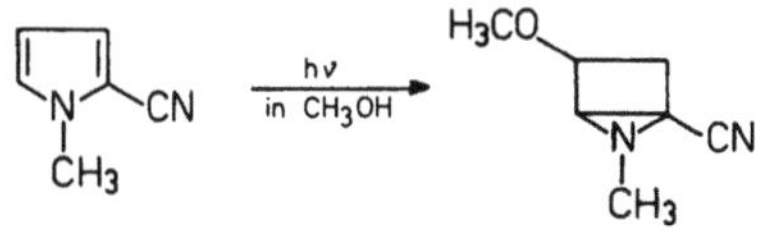

(113)

The structure of the methanol adduct was later corrected to be an isomer shown in Eq. 114. The Dewar compound was trapped by furan [115]. These reactions show that the reactivity of the 3-cyano isomer is higher than that of the 2-cyno isomer. Both authors did not isolate a Dewar isomer, but the obtained results suggest the intermediate occurrence of Dewar isomers and the formation of an equilibrium between them by a walk of the N—CH_3 part (114).

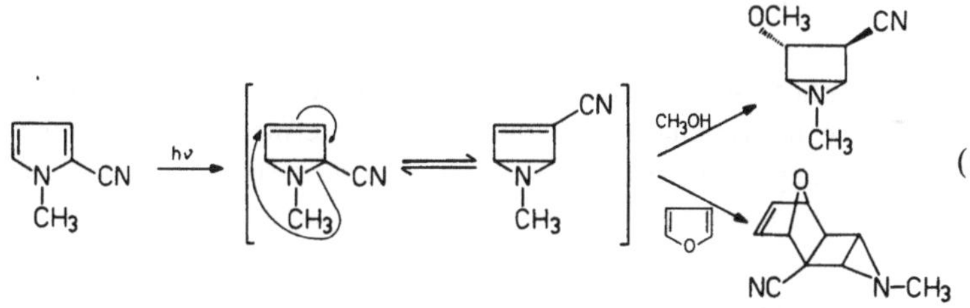

(114)

On the other hand, the photoreaction of 2-cyanofuran in methanol gives a methanol adduct of a cyclopropenyl aldehyde isomer (115) [114]:

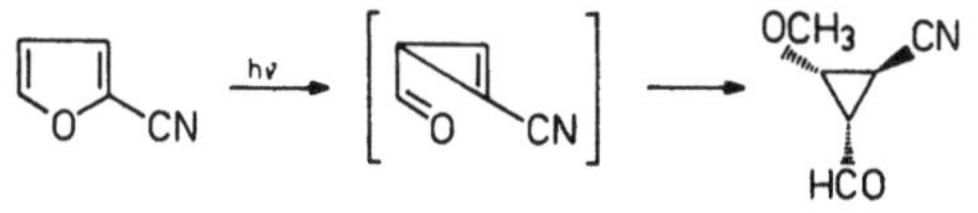

(115)

Depending on the formed intermediates, the photoreactions of five-membered heteroaromatic compounds may be classified into two types, these involving formation of cyclopropenyl ketones and Dewar intermediates. However, there are some reactions where both types of intermediates participate. Kojima and his coworker showed that both isomers should be taken into account in the photoreaction of an 1,3-oxazole compound (116) [116].

The photoreaction of 1,3-thiazoles was investigated extensively by Maeda and Kojima. They proposed tricyclic intermediates based on the take-up of deuterium

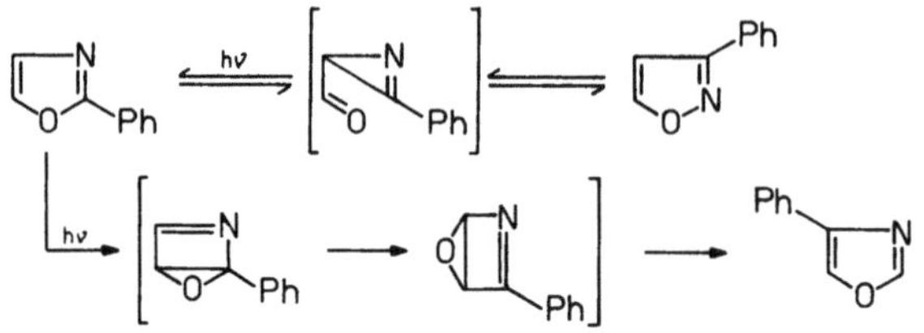

(116)

in the presence of deuterium oxide [117]. These intermediates are isomerized via Dewar intermediates to the corresponding 1,2-thiazoles (117).

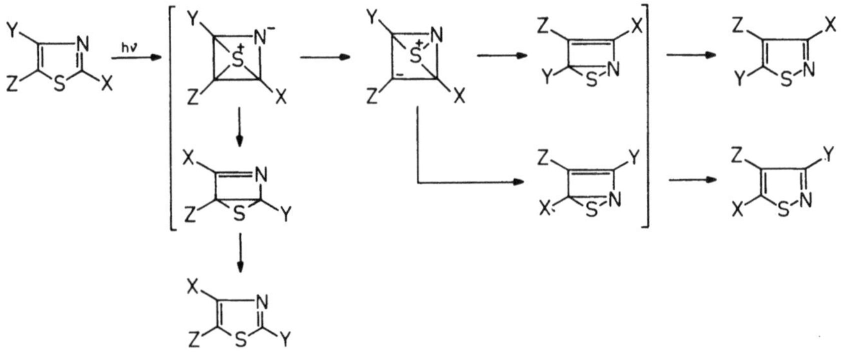

(117)

Thus, the photoisomerization of five-membered heterocycles seems not to be limited to one intermediate only but several intermediates may take part in this isomerization, depending on the nature of the ring system and substituents.

Mesoionic five-membered heterocycles are also an important group of heterocyclic compounds. These compounds are very useful intermediates in organic synthesis because they react as dipolar components in 1,3-dipolar cycloaddition reactions. They are photochemically isomerized to other dipolar compounds via Dewar-type intermediates. These dipolar compounds are converted to products different from those formed in the thermal reaction (118) [118]. (Concerning this type of reaction, see ref. [102].)

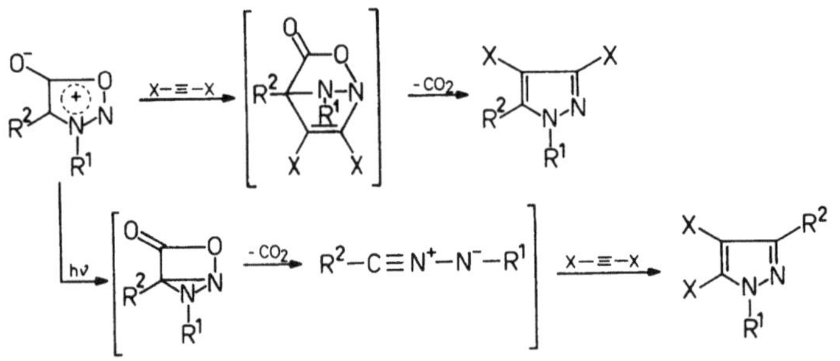

(118)

$R^1 = CH_3$, $R^2 = H$, $X = COOCH_3$

3.2 Reactions of Tetrakis(trifluoromethyl)Dewar thiophene

So far, the formation of valence-bond isomers by photolysis has mainly been discussed. Tetrakis(trifluoromethyl) Dewar thiophene is stable at room temperature and was used for the synthesis of Dewar pyrroles, a Dewar furan, and other interesting compounds from the standpoint of the structural chemistry. In this section, some of these reactions will be discussed.

As described in the previous section, the Dewar thiophene is a suitable dienophile and reacts with many kinds of dienes [119]. If the sulfur atom of the thiirane part can be eliminated, the Dewar thiophene will be a suitable precursor of a $C_4(CF_3)_4$ unit. An example is the synthesis of oxahomocubanes. Thus, the Diels-Alder adducts from the Dewar thiophene and furans were treated with triphenylphosphine and irradiated with a low-pressure mercury lamp to give oxahomocubanes (119). A peculiar point is the fact that *exo*-tricyclic compound is cyclized. Therefore, isomerization should have occurred prior to cyclization. This suggests that photocyclization cannot be used for the determination of such configurations [110].

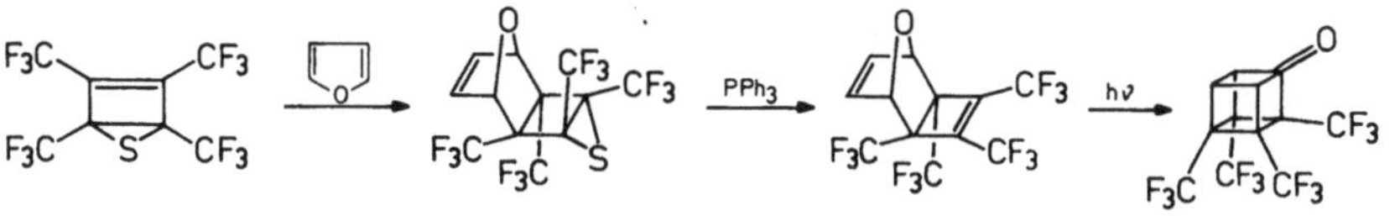

(119)

The 1,3-dipolar cycloaddition of azides to the Dewar thiophene proceeds smoothly. The obtained adducts are converted to the corresponding Dewar pyrroles. Some reactions of the adducts with azides are shown in scheme (120) [120–124].

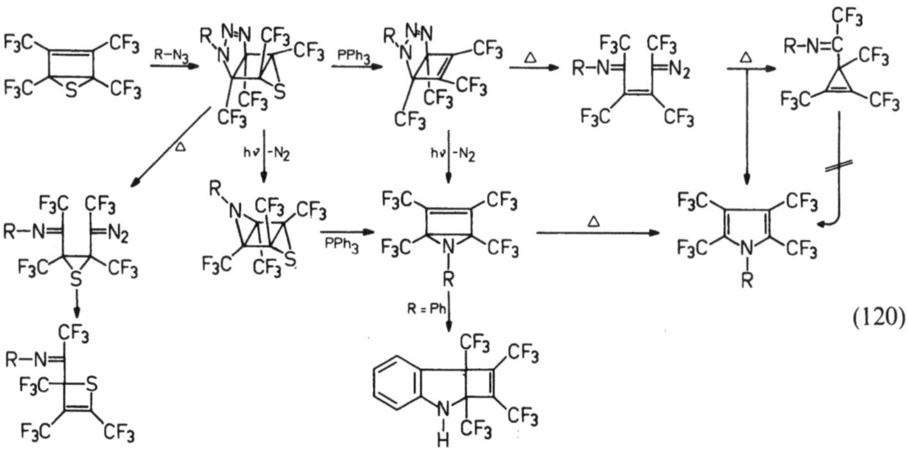

(120)

The most characteristic feature of the Dewar thiophene is that it reacts with all the azides examined even with hydrazoic acid [121]. Hydrazoic acid is reported to react with acetylenes but not with olefins. Therefore, the double bond of the Dewar thiophene is one of the most reactive olefinic double bonds.

Stepwise elimination of a sulfur atom and nitrogen molecule from the adduct gives the Dewar pyrroles [120, 121]. The stability of the desulfurized products depends on the substituents on the nitrogen. The N-phenyl derivative of the cyclobutatriazoline isomerizes spontaneously to the corresponding diazoimine which is converted into N-phenylpyrrole by the extrusion of a nitrogen molecule. Therefore, to obtain the N-phenyl Dewar pyrrole, denitrogenation must precede desulfurization while the N-cyclohexyl Dewar pyrrole is formed in both processes.

N-Phenyl Dewar pyrrole rearranges spontaneously to a cyclobutaindole while other Dewar pyrroles are stable at room temperature and thermally or photochemically isomerized to the pyrroles [124].

The thermolysis of the 1,3-dipolar adducts of tetrakis(trifluoromethyl)thiophene and azides gives thiethyltrifluoromethylketone imines through the diazocompound (see Scheme 120) [122]. In conclusion, the 1,2,3-triazoline moiety of the 1,3-dipolar adducts of the Dewar thiophene and azides or their desulfurized derivatives is denitrogenated by photolysis to give the aziridines, while it is thermally cleaved in a retro Diels-Alder manner to give diazoimine compounds (see Scheme 120). The high stability of the thietes may be attributed to the effect of trifluoromethyl groups.

The Diels-Alder adduct formed from the Dewar thiophene and pyrrole was used for the synthesis of a Dewar furan (121) [125].

The Dewar furan isomerizes not to the furan but to the corresponding cyclopropenyl ketone. This may be due to the weak oxygen-carbon bond of Dewar furan

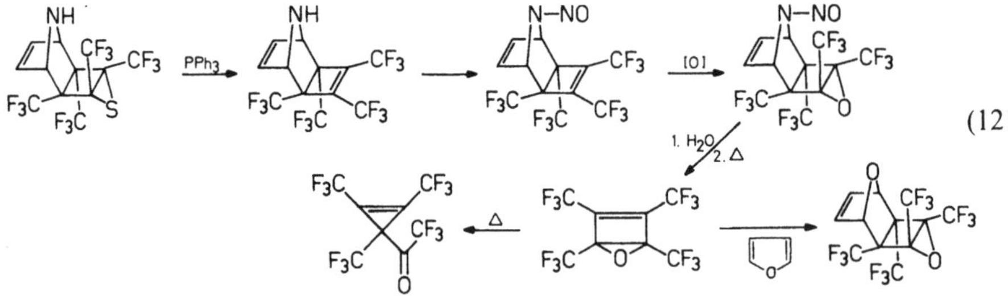

(121)

and the lower aromaticity of the furan ring. The Dewar furan is a more efficient dienophile than the Dewar thiophene. According to Lemal the difference in the reactivity of the two compounds may be attributed to the electronic effect of sulfur and oxygen. However, our results of the study of the Diels-Alder reactions suggest that steric effects play an important role [102].

Lemal et al. studies the 1,3-dipolar addition reaction of the Dewar thiophene with trifluoromethyl diazomethane. The cycloadduct shows reactions which are similar to those of the azide adducts (122) [126].

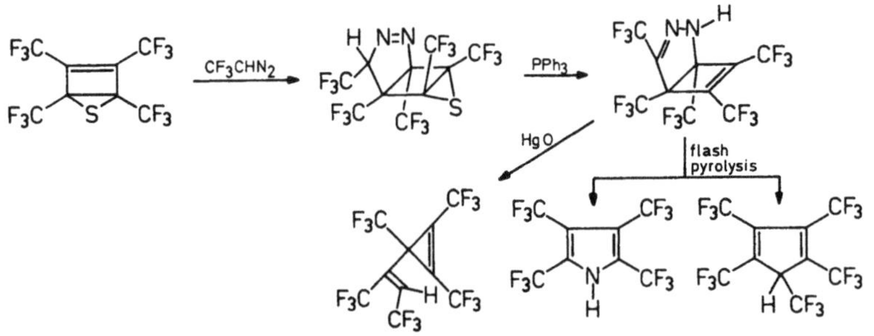

(122)

The Diels-Alder adduct of tetrakis(trifluoromethyl) Dewar thiophene and butadiene is transformed to tetrakis(trifluoromethyl)cyclooctatetraene [127] which is further converted to two semibullvalenes (123) [128].

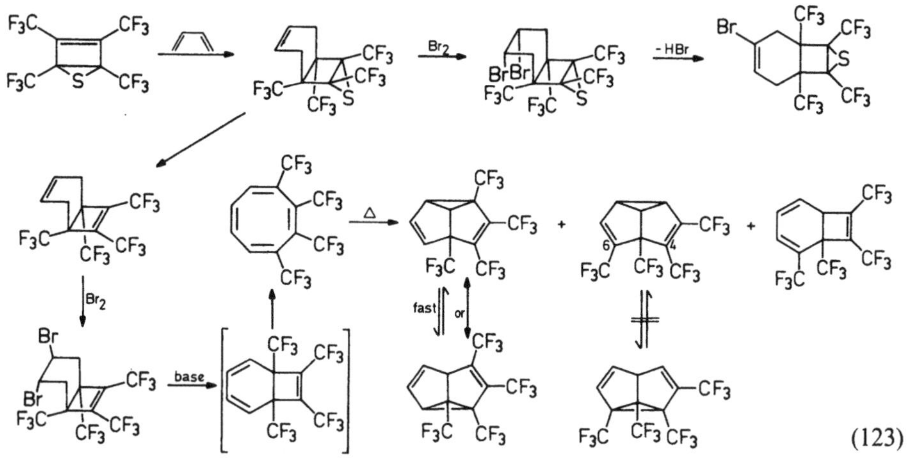

(123)

Bromination of the Diels-Alder adduct gives a cis-dibromide from which only one hydrogen bromide molecule is eliminated. This cis-addition may be due to the steric effect of the trifluoromethyl groups on the thiirane ring. This steric hindrance is removed by desulfurization leading to the trans-dibromide. The dibromide is converted to 1,2,3,8-tetrakis(trifluoromethyl)cyclooctatetraene. Photolysis of this cyclooctatetraene does not give a semibullvalene but its thermolysis yields two semibullvalenes [128]. The NMR spectra of these compounds show that the former semibullvalene contains two sp^2 and two sp^3 ring carbons, and the other four ring carbons are just between the regions of the sp^2 and sp^3 carbon atoms. This fact suggests that this compound is in a rapid equilibrium between the two Cope isomers. However, these signals are independent of the temperature. Thus, the activation energy for the Cope isomerization may be lowered by the substituent effect so that the transition state is frozen out. On the other hand, the latter semibullvalene is not involved in any equilibrium even at high temperature. This may be due to the repulsion between two trifluoromethyl groups at the 4- and 6-positions [128].

The thiirane part of the Dewar thiophene is not desulfurized by triphenylphosphine but isomerization to the thiophene occurs. This reaction with trivalent phosphorus compounds which are not substituted with electronegative groups is typical of Dewar thiophene. Therefore, a mechanism involving electron transfer from the phosphorous compound to the sulfur atom of the Dewar thiophene has been proposed. The reaction of Dewar thiophene with diphenylchlorophosphine proceeds via an intermediate which was isolated (124) [129].

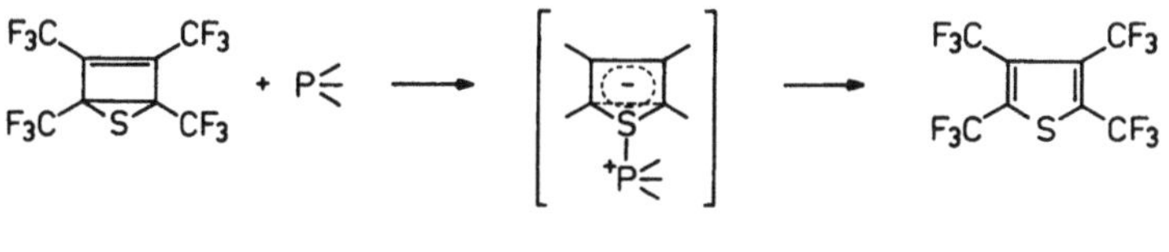

(124)

On the other hand, Lemal and his coworkers prepared Dewar thiophene sulfoxide. A study of the NMR spectrum reveals that the sulfoxide group migrates around the cyclobutene ring [130]. They called this migration a pseudopericyclic reaction. An investigation of the temperature-dependent NMR spectrum shows that this migration takes place on the Dewar thiophene itself (125) [131].

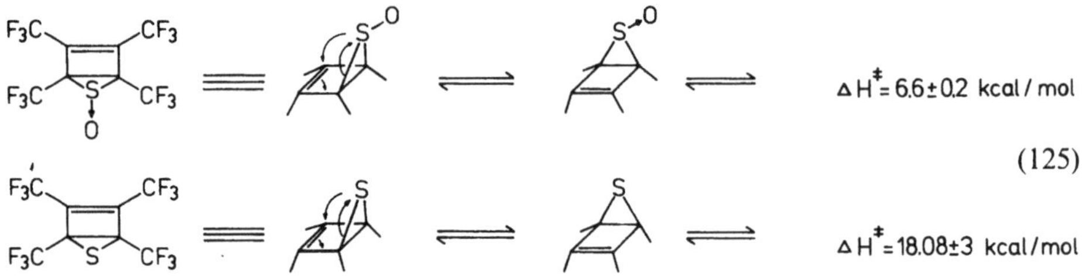

(125)

This mechanism was reinvestigated by Snyder's group using molecular orbital calculations [132]. They suggested that the intervention of the lone-pair electrons is not essential. Thus, this reaction would be a simple pericyclic reaction.

3.3 Six-Membered Heterocyclic Compounds

There are many kinds of six-membered heterocyclic compounds but only few valence-bond isomers have been isolated. Thus, we will first describe the photoreactions of pyrones and pyridones and subsequently those of the usual heteroaromatic compounds with increasing number of hetero atom's.

Corey and Streith found that the photolysis of pyrone and pyridone yields bicyclo[2.2.0]hexane analogs [133]. They suggested that the product has a structure formed by formal addition of carbon dioxide to cyclobutadiene (126).

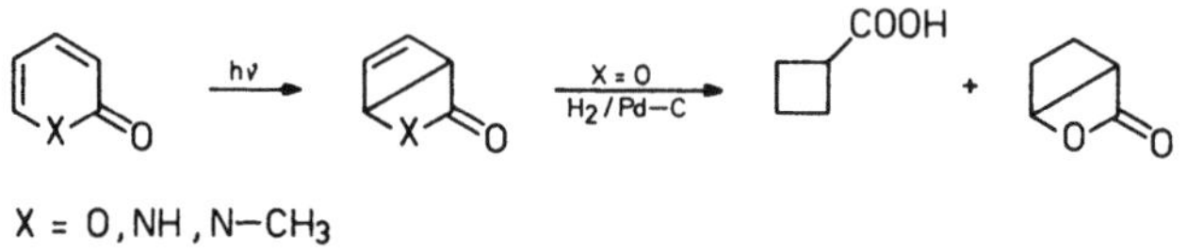

(126)

The reverse reaction has also been observed: Irradiation of the Dewar α-pyrone in an argon atmosphere affords cyclobutadiene which is dimerized to tricyclo-[4.2.0.0^{2,5}]octadiene upon heating or photolyzed to acetylene (127) [134,135].

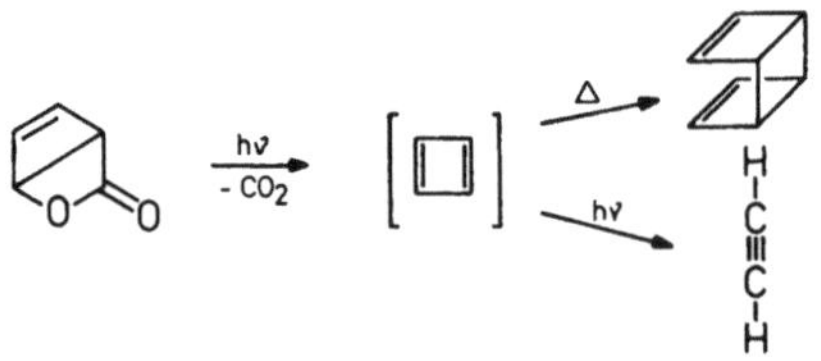

(127)

The assignment of the IR spectrum of cyclobutadiene was later corrected [136] and the above photochemical process was investigated in detail by means of isotope labelling (128) [137].

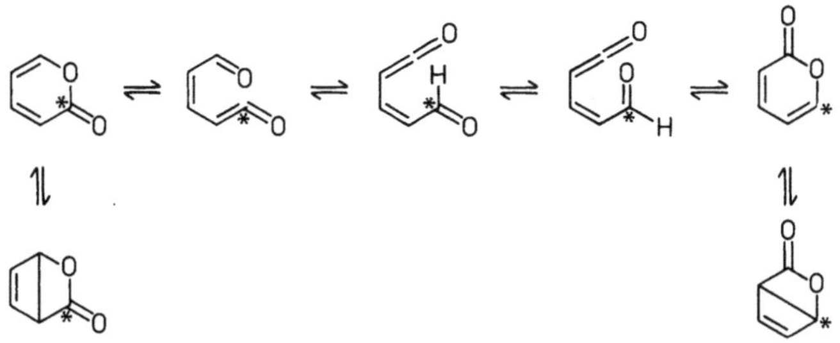

(128)

The application of this reaction to an aza analog leads to an azacyclobutadiene [138]. This compound is cleaved to two 1-alkyne and two nitrile molecules which suggests that an equilibrium exists between the two azacyclobutadiene structures (129). This reaction is widely used for the synthesis of cyclobutadiene intermediates.

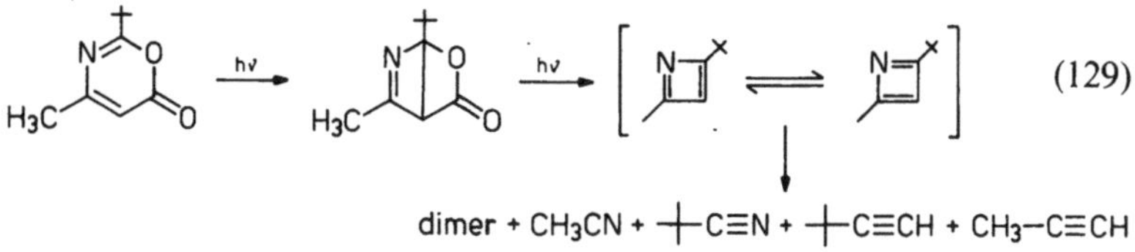

(129)

The photoreaction of concentrated pyridone solutions usually gives dimers whereas in dilute solution, an isomer of the Dewar type is formed (130) [139].

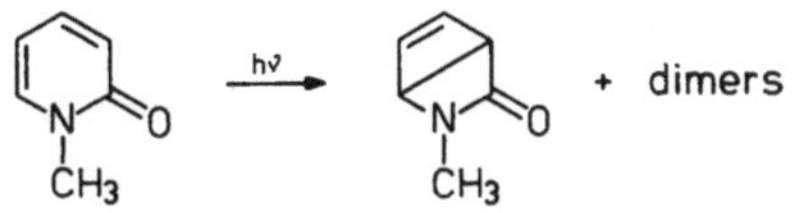

(130)

The photoisomerization of N-methyl-α-pyridone has widely been studied. One of its applications is illustrated by the synthesis of β-lactam derivatives (131) [140].

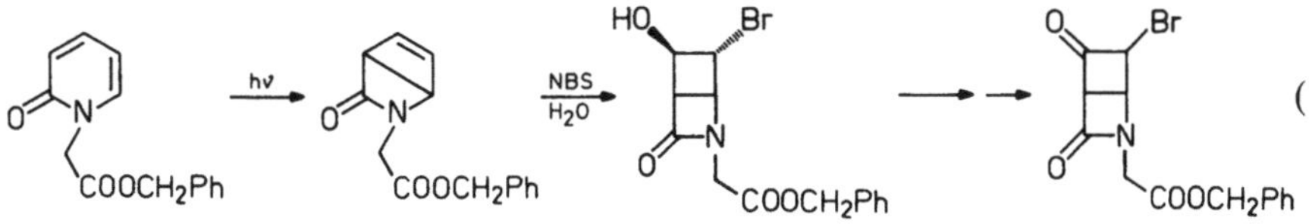

(131)

Kaneko and his coworkers reported an isomerization proceeding via a β-lactam isomer (132) [141].

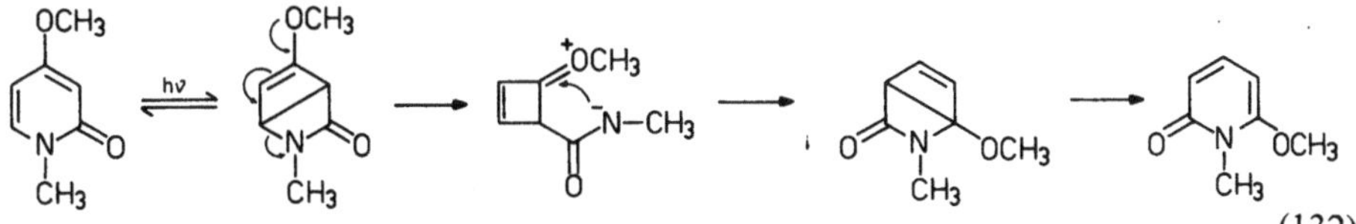

(132)

This type of Dewar compound was formed from a cyclobutadiene derivative [142]. This process may be regarded as the retroreaction of cyclobutadiene formation (133).

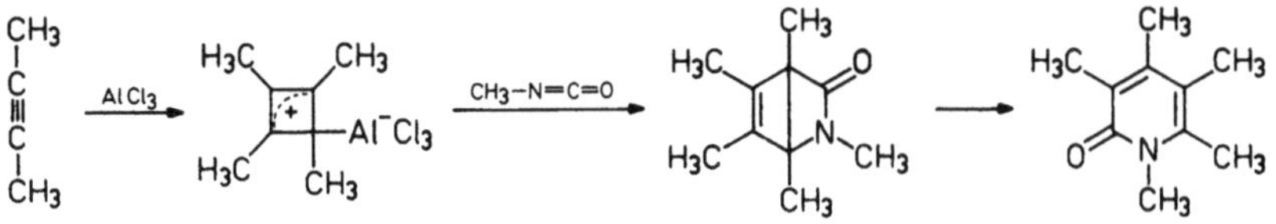

(133)

The photoreaction of a mesoionic compound yields a 2,6-bonded isomer together with a dimer (134) [143].

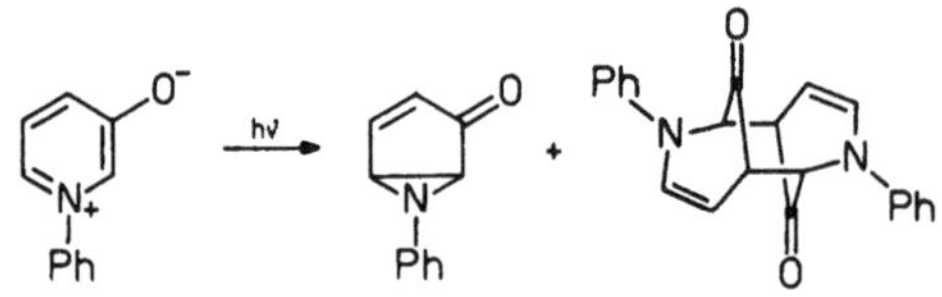

(134)

1,2-Dihydropyridine is fairly unstable and susceptible to oxididation but the Dewar isomer prepared from it by photolysis is rather stable and used as masked 1,2-dihydropyridine (135) [144].

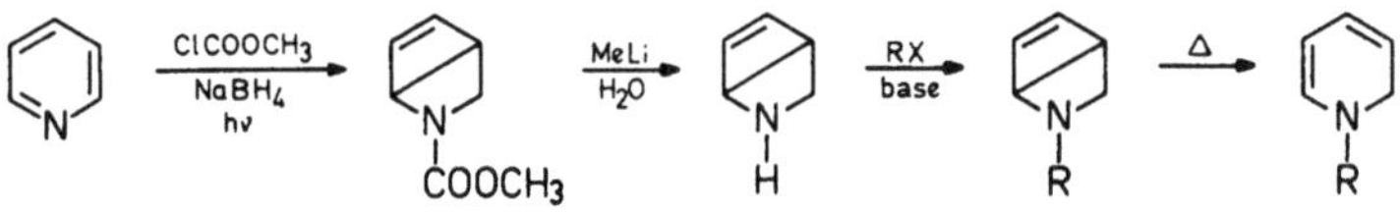

(135)

Thermolysis of N-alkyl-1,2-dihydropyridines causes ring opening and Cope-type recyclization to give the corresponding dihydropyridine isomers (136) [145].

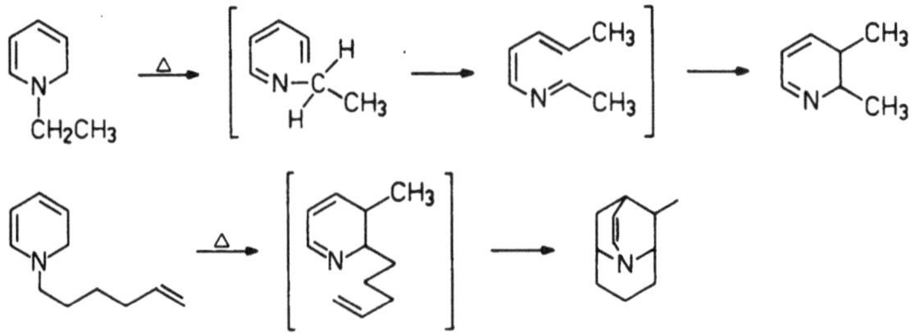

(136)

Another type of ring transformation occurs when the pyridine bears an electron-attracting substituent (137) [146].

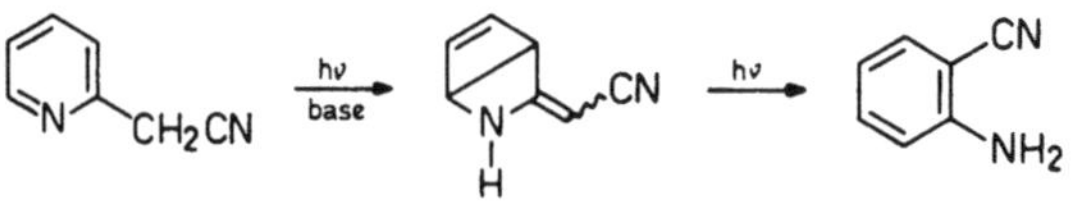

(137)

The cyclobutene part of the Dewar-type isomer reacts as a dienophile with tetraphenylcyclopentadienone. The resulting product is photolyzed to an azacyclobutene (138) [147].

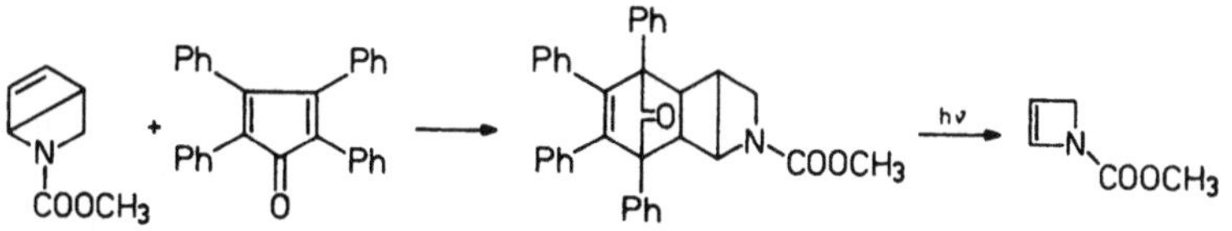

(138)

The photolysis of substituted pyridines causes migration of the substituent, and a mechanism involving intermediate formation of a Dewar-type isomer and an azaprismane has been proposed [148]. However, the formation of these intermediates has not been confirmed so far (139).

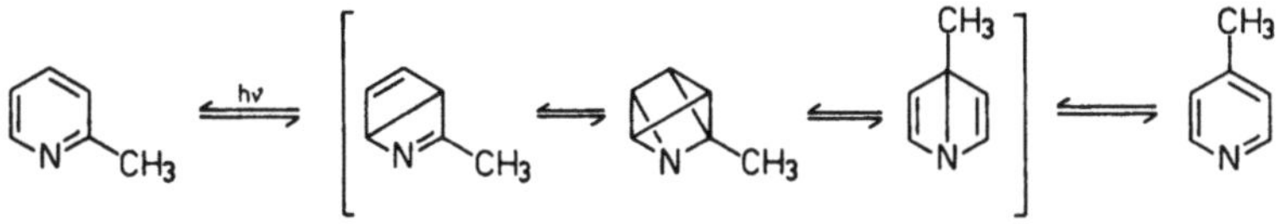

(139)

On the other hand, Wilzbach detected 2,5-bonded Dewar pyridine by UV spectrometry as the photolysis product of pyridine and isolated 1,2-dihydro-2,5-bonded Dewar pyridine after reduction of the reaction mixture with sodium borohydride (140) [149]. This result is in accordance with MO calculations according to which a 2,5-bonded Dewar pyridine is expected to be more stable than the 1,4-bonded isomer (141) [150].

(140)

(141)

The unsubstituted Dewar pyridine itself is too unstable to be isolated in a pure form. Haszeldine and his coworkers reported that the photolysis of perfluoro-(pentaethylpyridine) yields a 1,4-bonded Dewar pyridine which is quite stable at room temperature and converted to azaprismane on further irradiation (142) [151].

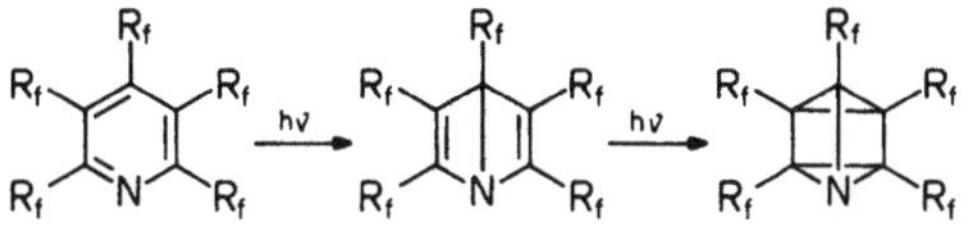

(142)

This is a further example of the stabilizing effect of perfluoroalkyl groups on a strained ring system. The fact that the 1,4-bonded isomer instead of the 2,5-bonded isomer is obtained may be explained by the repulsions between the perfluoroethyl groups on the bridge heads of the 2,5-bonded isomer which make this isomer less stable than the 1,4-bonded isomer.

Later, another 1,4-bonded Dewar pyridine bearing methyl and trifluoromethyl groups in alternating positions was obtained [152]. This compound is fairly stable but much less stable than the perfluoro(pentaethyl) isomer. The stability of the former isomer seems to be partly due to the electronic pull-push interactions between the substituents on the double bond. A large coupling is observed between the methyl and the trifluoromethyl groups on the double bond in the ^1H- and ^{19}F-spectra but a singlet of the 4-methyl group. Furthermore, the protons of the 2,6-methyl groups of 3,5-bis(trifluoromethyl)-2,4,6-trimethyl-1,4-bonded Dewar pyridine are deuterated by treatment with sodium deuteromethoxide in deuteromethanol whereas the protons of the 4-methyl group are not replaced by deuterium (143) [153].

The isomerization of this Dewar pyridine to the aromatic pyridine is catalyzed by acids and some heavy metals. Palladium and platinum complexes of this Dewar pyridine have been isolated and analyzed by X-ray analysis [152]. The kinetics of this isomerization was studied using the complexes [154].

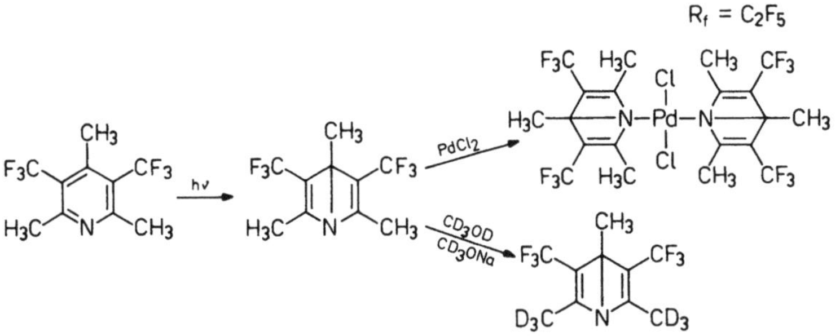

(143)

The above two Dewar pyridines bear the same substituents in the 2-, 4-, and 6-position. Therefore, isomerization between these positions could not be observed. Chambers and his coworkers examined the photoreaction of a perfluoropyridine bearing three kinds of substituents and suggested that first a 2,5-bonded Dewar pyridine is formed which is subsequently isomerized to the 1,4-bonded Dewar compound although the 2,5-bonded isomer has not been observed (144) [155].

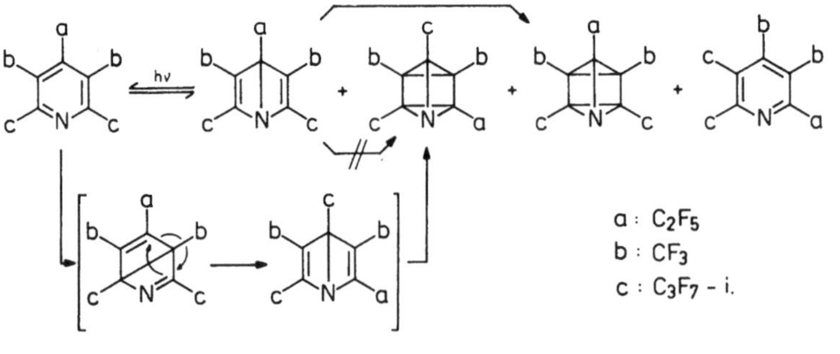

(144)

In 1977, the 2,5-bonded Dewar isomers of perfluoroalkylated pyridine were isolated by Chamber's group (145)[156]. The photolysis was carried out in the gas phase. Therefore, thse isomers must be the primary products. Their distributions seems to be determined by the electronic interaction between the substituents on the double bonds and steric effects.

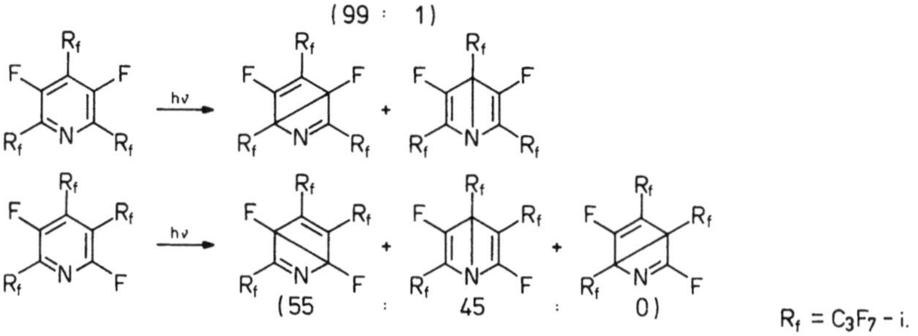

(145)

Next, some reactions of heterocyclic compounds containing two nitrogen atoms will be discussed. First, 4,6-dimethyl-1,2-dihydropyrimidone was isomerized to a bicyclic compound which was degraded to 2,4-dimethylazacyclobutadiene (146)[157].

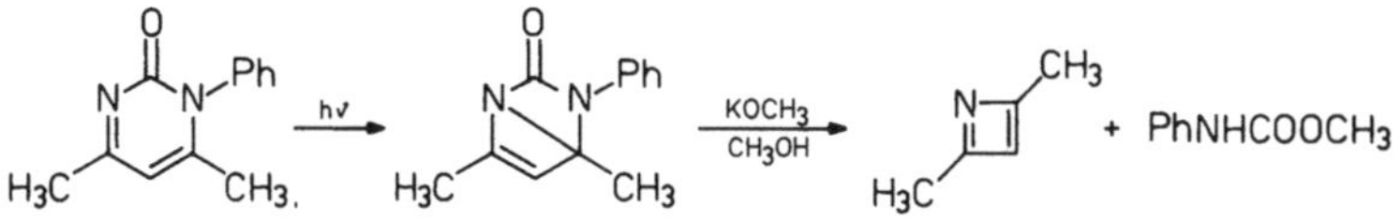

(146)

The cycloaddition reaction of cyclobutadienes with azodicarboxylates gives 3,6-bonded Dewar pyridazines. An example of this type of reaction has already been reported before[94]. A further example is described by the reaction sequence (147)[158,159]. The formed cycloadduct a hydrolyzed, decarboxylated, and oxidized to 3,6-bonded Dewar pyridazine which loses nitrogen to form cyclobutadiene (147).

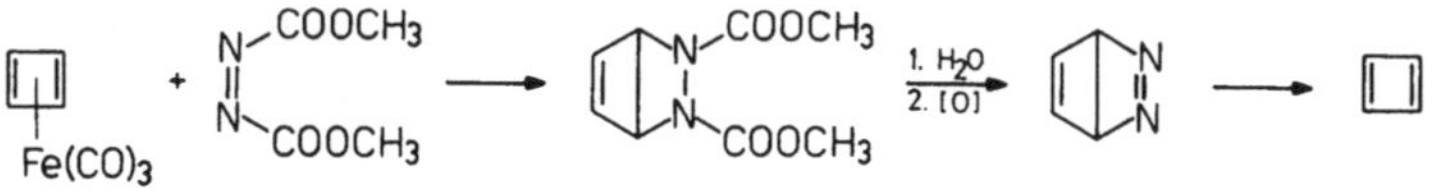

(147)

The same bicyclic compound is obtained by photolysis of the monocyclic isomer (148)[160].

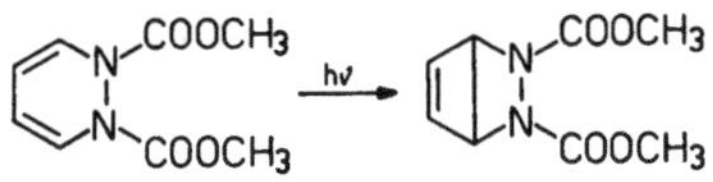

(148)

1,2-Methoxycarbonyl-3,6-bonded Dewar pyridazine reacts as a dienophile. This reaction is used for the synthesis of dimethyl 1,2-diazacyclobut-3-ene-1,2-dicarboxylate. This compound has two pairs of unshared electrons but does not show any aromaticity (149) [161].

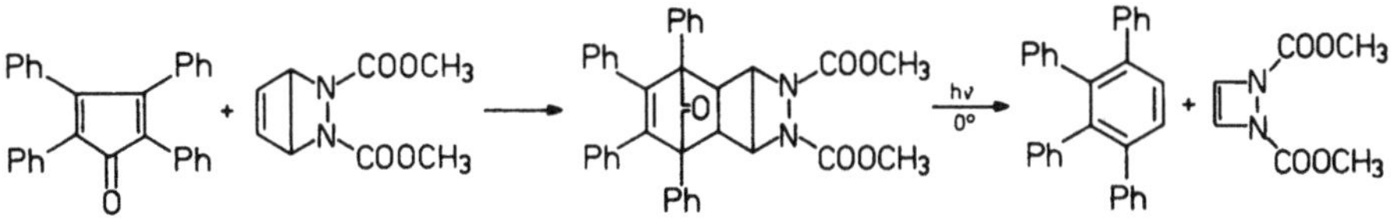

(149)

Valence-bond isomers of perfluorinated diazine derivatives have been extensively studied by Chambers' group.

Tetrafluoropyridazine is photochemically isomerized to tetrafluoropyrazine and thermally to tetrafluoropyrimidine. These reactions are proposed to proceed via the corresponding Dewar and the prismane analogs, and benzvalene analog, respectively (150) [162].

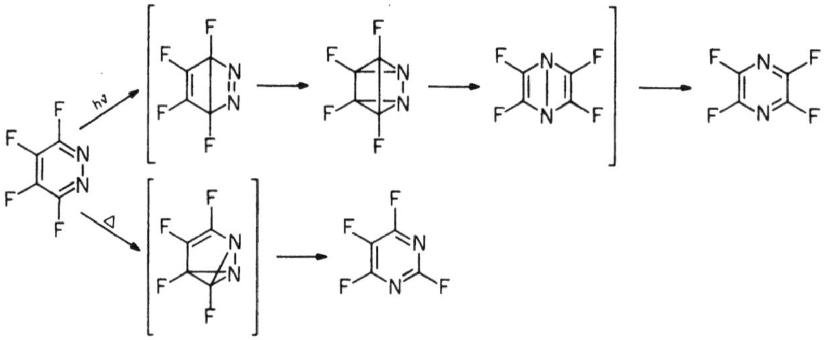

(150)

If this mechanism applied to perfluoro(4,5-diisopropylpyridazine), perfluoro(2,3-diisopropylpyrazine) should be obtained. However, the 2,5-substituted isomer is obtained together with its Dewar isomer [163]. In a more detailed study, a 2,5-bonded Dewar pyridazine was isolated which isomerized to the Dewar pyrazine. Therefore, another mechanism involving isomerization of the Dewar pyridazine to the Dewar pyrazine which does not proceed via the prismane has been proposed (151) [164].

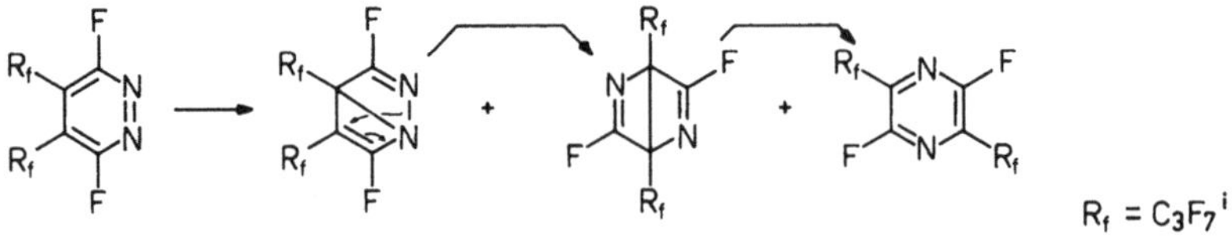

(151)

Furthermore, the photolysis of perfluoro(3,5-diisopropylpyridazine) yields unusual products, azacyclobutadiene dimers which seem to be formed through photolysis of the Dewar compound. On the basis of this result a mechanism of the photoisomerization of the perfluoropyridazines was established (152) [165].

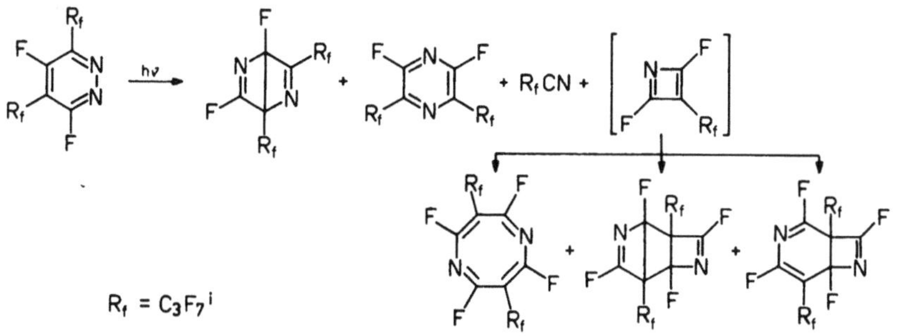

(152)

The thermolysis of perfluorinated pyridazines reveals that the isomerization of perfluoropyridazine is strongly accelerated by the addition of perfluoro(4,5-di-isobutylpyridazine) which isomerizes very rapidly. Therefore, once Chambers' group proposed a thermally sensitized mechanism [166] but later, they corrected the mechanism to be a radical chain mechanism (153) [167]. Perfluoropyridazine does not react at 300 °C, but it reacts by the addition of *2* to give 21% of *3a*. Addition of $CF_3CF(C_2F_5)$—$CF(CF_3)C_2F_5$ also accelerates this reaction.

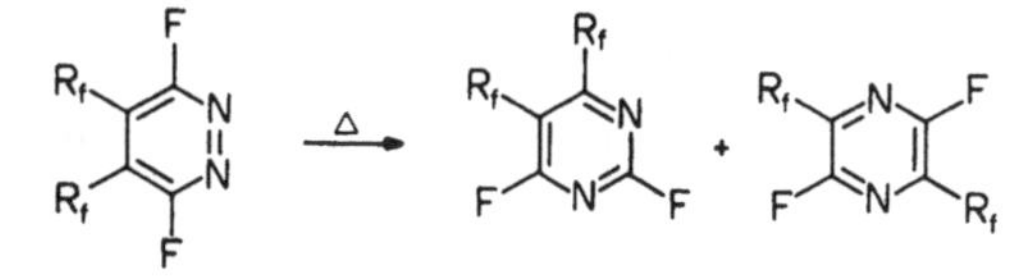

R :	$C_3F_7{}^i$	1	1a	
	$CF(CF_3)C_2F_5$	2	2a	2b
	F	3	3a	

Mechanism.

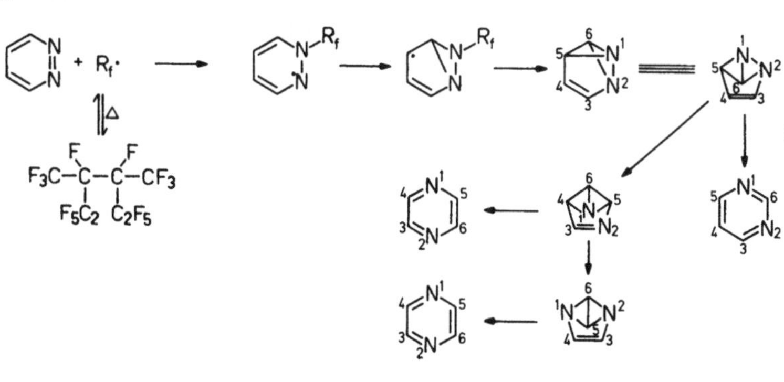

(153)

To conclude, it can be said that the photoreaction of perfluorinated pyridazines proceeds via 1,4-bonded Dewar pyridazines while their thermolysis occurs via the benzvalene analogs which are generated by a radical mechanism.

The thermolysis of perfluoro(triisopropyl-1,2,4-triazine) gives an alkyne and a nitrile (154) [168,169].

The photolysis of the same compound yields an s-triazine, alkyne and a nitrile (155) [168].

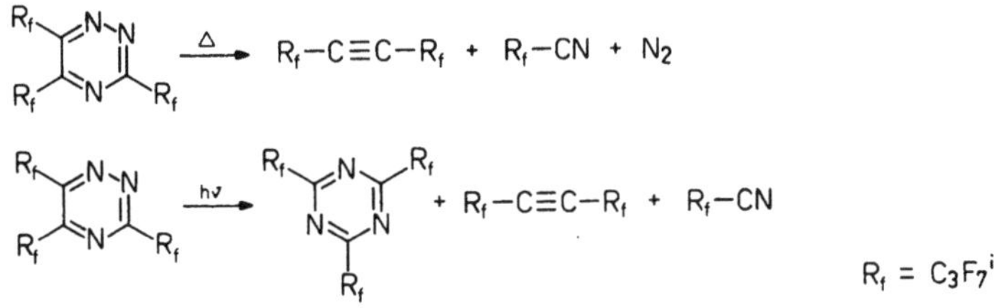

(154)

(155)

Gompper and his coworkers treated tris(dimethylamino)cyclopropenium cation with potassium azide and obtained a 1,2,3-triazine. They assumed that this reaction proceeds via a benzvalene-type intermediate (156) [170]:

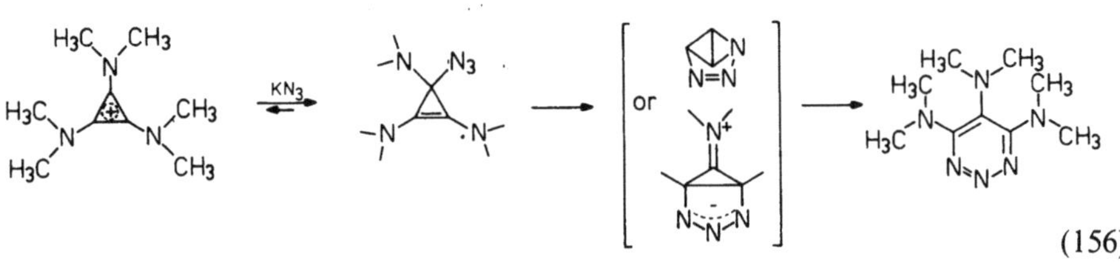

(156)

So far, valence bond isomers of oxygen-, sulfur- and nitrogen-containing heteroaromatic compounds were discussed where no benzvalene-type isomers had been isolated. The authors of this article synthesized tetrakis(trifluoromethyl)-1,4-diphosphabenzene [171] and obtained a diphosphabenzvalene by the photoreaction [172]. The trifluoromethyl groups stabilize diphosphabenzene which is an electronically unstable compound (157). Thus, tetrakis(trifluoromethyl)-1,4-diphosphabenzene can be isolated.

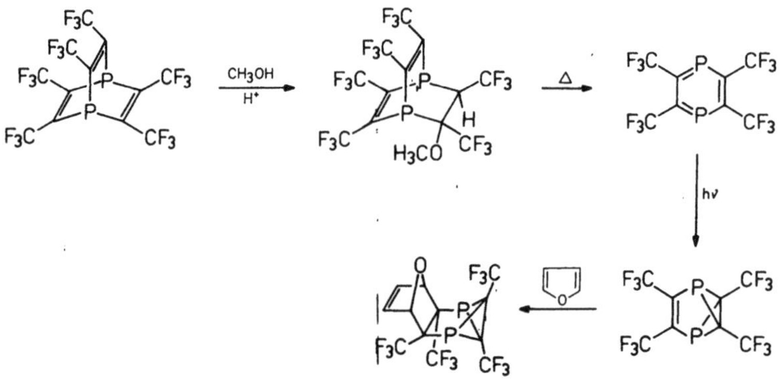

(157)

3.4 Seven Membered Heterocyclic Compounds

Seven-membered heterocyclic compounds such as oxepin or azepin are not considered to be aromatic, and the number of their valence bond isomers is so large that we cannot describe all these compounds. Therefore, only a few interesting examples will be given.

In the following two examples a five-membered ring is converted into a bicyclo[3.2.0]-heptadiene system on treatment with dimethyl acetylenedicarboxylate (158, 159).

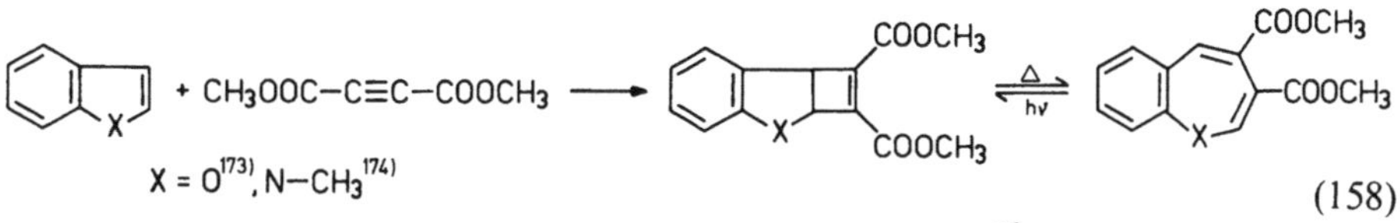

(158)

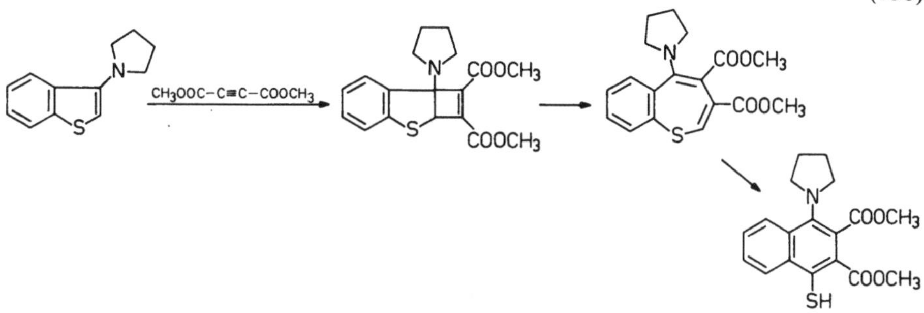

(159)

In these examples the electron-rich olefinic double bond reacts with the electron-deficient triple bond of the acetylenedicarboxylate. Further examples in which a five-membered ring is converted into the corresponding bicyclic system are reactions (160) and (161). Here, the electron-deficient olefinic double bond of five-membered heterocycle reacts with the electron-rich triple bond of the alkyne. The latter example includes the synthesis of oxahomoprismane.

Valence-bond isomers having a bicyclobutane part have been synthesized only when they are fused with a benzene ring (162, 163) [178, 179]. However, due to the stabilizing effect of trifluoromethyl groups non-fused compounds have been isolated.

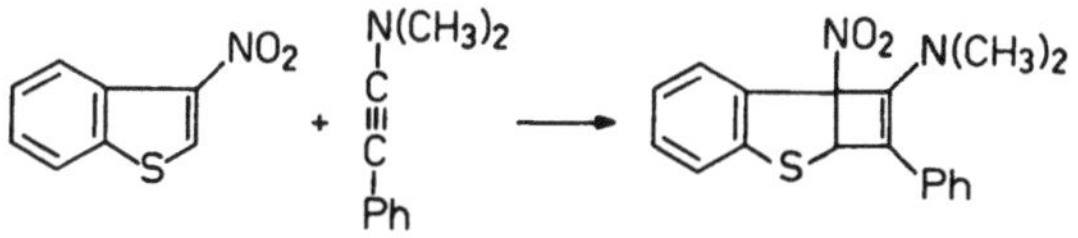

(160) [176]

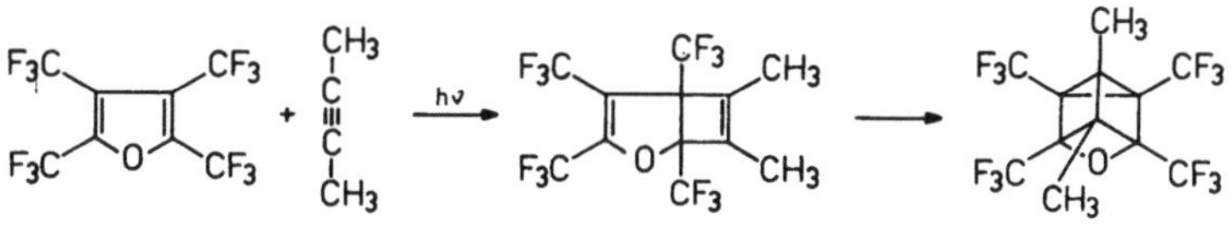

(161) [177]

Thus the valence-bond isomers of hexakis(trifluoromethyl)oxepin are formed from the benzvaleneozonide, as shown before [100]. The photoreaction of the tricyclic

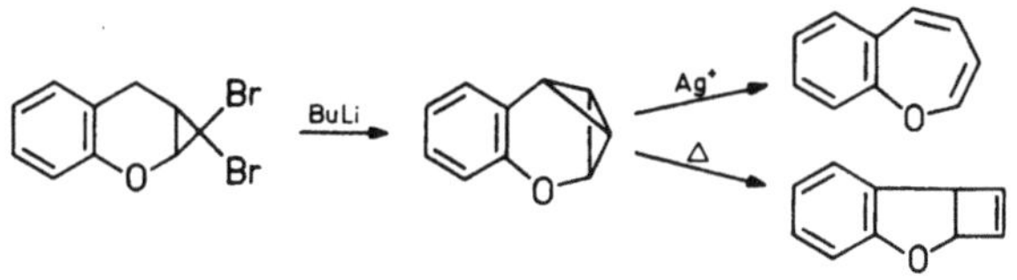

(162)

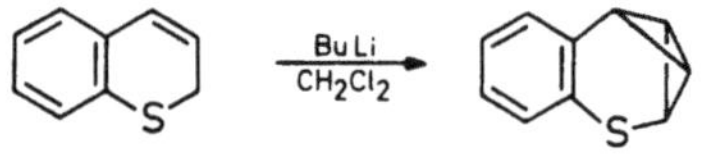

(163)

isomer gives two cyclopropene compounds the *cis*-isomer of which is thermally retransformed to the starting compound. Both isomers are photochemically converted into a cyclopropenyloxetene. The high stability of this compound is remarkable (164) [180].

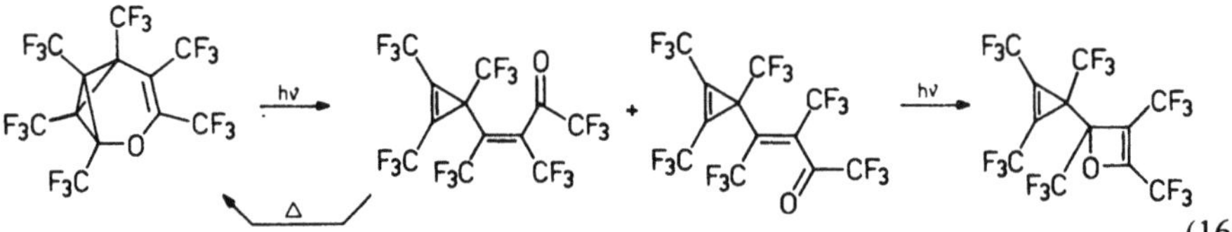

(164)

The thermal back reaction to the tricyclic compound has attracted our attention. The application of this reaction to a cyclopropenyl ketone imine system gives a similar isomer of a diazepin compound (165) [181]. This is the first synthesis of a valence-bond isomer of a nitrogen-containing heterocyclic compound having a bicyclobutane part.

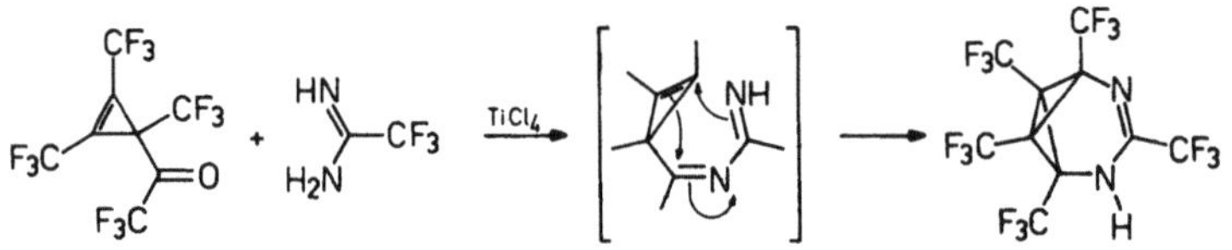

(165)

Azepins or 1,3-diazepins are formed by photolysis of N-ylides. These compounds are further photoisomerized to [3.2.0] ring systems. Details of these reactions are sumarized by Tsuchiya [182]. Some of them are described in schemes (166) and (167) [183, 184].

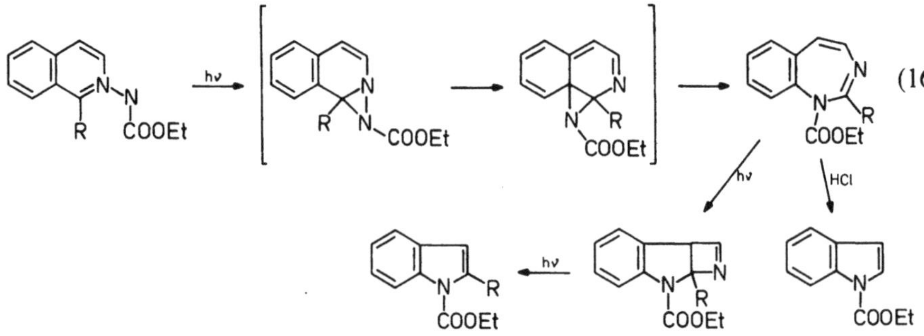

(166)

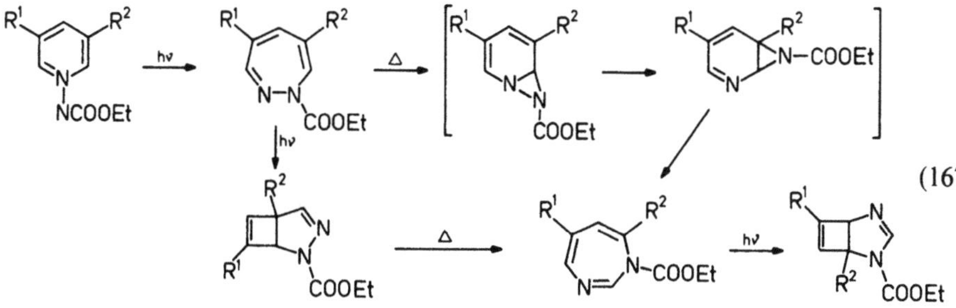

(167)

4 Concluding Remarks

So far, we have shown synthesis and reactions of valence bond isomers of aromatic compounds, and much less of theoretical parts are discussed. However, these compounds have extensively been studied by means of the molecular orbital theory. Some MO calculations are used for the interpretation of photoelectron spectra.

About 25 years ago, most of the valence-bond isomers described in this review were beyond the dream of chemists but meanwhile these isomers have been synthesized. Some of them are expected to be used for the utilization of the solar energy. Most of these highly reactive compounds are very useful for the synthesis of other compounds.

5 References

1. Baker, W., Rouvray, D. H.: J. Chem. Educ. *55*, 645 (1978)
2. van Tamelen, E. E., Pappas, S. P.: J. Am. Chem. Soc. *84*, 3789 (1962)
3. Viehe, H. G. et al.: Angew. Chem. *76*, 922 (1964)
4. van Tamelen, E. E., Pappas, S. P.: J. Am. Chem. Soc. *85*, 3297 (1963)
5. Ward, H. R., Wishnock, J. S.: ibid. *90*, 1086 (1968)
6. Criegee, R., Zanker, F.: Angew. Chem. *76*, 716 (1964)
7. Watts, L., Fitzpatrick, J. D., Pettit, R.: J. Am. Chem. Soc. *87*, 3253 (1965)
8. Maier, G., Schneider, K. A.: Angew. Chem. *92*, 1056 (1980)
9. Schäfer, W., Hellman, H.: ibid. *79*, 566 (1967)
10. Vatanatham, S., Farona, M. F.: J. Catal. *61*, 540 (1980)
11. Kammula, S. L. et al.: J. Am. Chem. Soc. *99*, 5817 (1977)
12. Weinges, K. et al.: Chem. Ber. *110*, 2961 (1977)
13. Camaggi, G., Gozzo, F.: J. Chem. Soc. (C) 1969, 489
14. Barlow, M. G., Haszeldine, R. N., Hubbard, R.: J. Chem. Soc., Chem. Commun. 1969, 202
15. Lemal, D. M., Staros, J. U., Austel, A.: J. Am. Chem. Soc. *91*, 3373 (1969)
16. Clifton F. D., Flowers, W. T., Haszeldine, R. N.: J. Chem. Soc., Chem. Commun. *1969*, 1216; Dabbagh, A. M. et al.: J. Chem. Soc. Perkin Trans. II *1979*, 1407
17. Semeluk, G. P., Stevens, R. D. S.: J. Chem. Soc. (D), *1970*, 1720
18. Brovko, V. V. et al.: Izv. Sibirsk. Otd. Akad. Nauk. SSR, Ser. khim. Nauk. *1971*, 103
19. Barlow, M. G., Haszeldine, R. N., Kershaw, M. J.: J. Chem. Soc. Perkin Trans. *I*, *1974*, 1736
20. Barlow, M. G., Haszeldine, R. N., Kershaw, M. J.: ibid. *1975*, 2005
21. Barlow, M. G., Haszeldine, R. N., Kershaw, M. J.: Tetrahedron *31*, 1649 (1975)
22. Delyagina, N. I. et al.: Dokl. Akad. Nauk. SSR *217*, 836 (1974), Tetrahedron *30*, 4031 (1974)
23. Goldberg, I. B., Crowe, H. R., Franck, R. W.: J. Am. Chem. Soc. *98*, 7641 (1976)
24. Yoshida, Z. et al.: Kokagaku Toronkai Yoshishu *1979*, 6; CA. *92*, 197660 (1980)

25. Guesten, H., Mintas, M., Klasinc, L.: J. Am. Chem. Soc. *102*, 7936 (1980); Jahn, B., Dreeskamp, H.: Z. Naturforsch. *37A*, 376 (1982)
26. Carty, D. T.: Tetrahedron Lett. *1969*, 4753
27. Pritschins, W., Grimme, W.: ibid. *23*, 1151 (1982)
28. Weinges, K. et al.: Angew. Chem. *93*, 1008 (1981)
29. Bryce-Smith, D., Gilbert, A.: Tetrahedron *32*, 1309 (1976); *33*, 2459 (1977)
30. Wilzbach, K. E., Kaplan, L.: J. Am. Chem. Soc. *87*, 4006 (1965)
31. Wilzbach, K. E., Ritscher, J. S., Kaplan, L.: ibid. *89*, 1031 (1967)
32. Katz, T. J., Wang, E. J., Acton, N.: ibid. *93*, 3782 (1971)
33. Burger, U., Mazenod, F.: Tetrahedron Lett. *1976*, 2881
34. Burger, U., Gandillon, G.: ibid. *1979*, 4281
35. Burger, U., Gandillon, G., Mareda, J.: Helv. Chim. Acta *64*, 844 (1981)
36. Burger, U.: Chimia *33*, 147 (1979)
37. Gandillon, G., Bianco, B., Burger, U.: Tetrahedron Lett. *22*, 51 (1981)
38. Schäfer, W.: Angew. Chem. Intern. Ed. Engl. *6*, 78 (1967)
39. Katz, T. J., Acton, N.: J. Am. Chem. Soc. *95*, 2738 (1973)
40. Woodward, R. B., Hoffmann, R.: The Conservation of Orbital Symmetry, Weinheim, Verlag Chemie/Academic Press 1970
41. Dewar, M. J. S., Ford, G. P., Rzepa, H. S.: J. Chem. Soc., Chem. Commun. *1977*, 728
42. Dewar, M. J. S., Kirschner, S.: J. Am. Chem. Soc. *96*, 7579 (1974)
43. Turro, N. J. et al.: Tetrahedron Lett. *1976*, 4133
44. LeNoble, W. J. et al.: J. Am. Chem. Soc. *104*, 3150 (1982)
45. Taylor, G. N.: Z. Phys. Chem. (Frankfurt/M) *101*, 237 (1976)
46. Jones II, G., Chiang, S.-H.: J. Am. Chem. Soc. *101*, 7421 (1979)
47. Jones II, G., Becker, W. G.: ibid. *103*, 4630 (1981)
48. Harman, P. J., David, W. T.: J. Phys. Chem. *85*, 2731 (1981)
49. Breslow, R., Khanna, P. L.: Tetrahedron Lett. *1977*, 3429
50. Lemal, D. M., Dunlap, Jr., L. H.: J. Am. Chem. Soc. *94*, 6562 (1972)
51. Sztuba, B., Ratajczak, E.: J. Chem. Soc., Perkin Trans. II *1982*, 823
52. Volger, H. C., Hogeveen, H.: Rec. Trav. Chim. Pays-Bas *86*, 1356 (1967)
53. van Tamelen, E. E., Carty, D.: J. Am. Chem. Soc. *89*, 3922 (1967)
54. Rondan, N. G. et al.: ibid. *103*, 236 (1981)
55. Hogeveen, H. et al.: ibid. *96*, 7518 (1974)
56. Heldeweg, R. F., Hogeveen, H., Zwart, L.: Tetrahedron Lett. *1977*, 2535
57. Hasselmann, D., Loosen, K.: Angew. Chem. *90*, 641 (1978)
58. Brune, H. A., Jobst, H., Lach, P.: Chem. Ber. *113*, 511 (1980)
59. Wight, F. R.: Tetrahedron Lett. *1976*, 3691
60. Carless, H. A. J., Trivedi, H. S.: J. Chem. Soc., Chem. Commun. *1981*, 950
61. Toy, M. S., Stringham, R. S.: J. Fluorine Chem. *13*, 23 (1979)
62. Toy, M. S., Stringham, R. S.: J. Org. Chem. *44*, 2813 (1979)
63. Barlow, M. G. et al.: J. Chem. Soc., Perkin Trans. I, *1975*, 2010
64. Kobayashi, Y. et al.: ibid. *1979*, 2253
65. Barlow, M. G., Haszeldine, R. N., Peck, C. J.: J. Fluorine Chem. *18*, 601 (1981)
66. Barlow, M. G., Haszeldine, R. N., Peck, C. J.: J. Chem. Soc., Chem. Commun. *1980*, 158
67. Gerace, M. J., Lemal, D. M., Ertl, H.: J. Am. Chem. Soc. *97*, 5584 (1975)
68. Burger, U., Mazenod, F.: Tetrahedron Lett. *1976*, 2885
69. Berger, U., Mazenod, F.: *1977*, 1757
70. Hutchings, M. G. et al.: J. Am. Soc. *99*, 7126 (1977)
71. Turro, N. J. et al.: J. Org. Chem. *42*, 92 (1977)
72. Aumann, R., Wörmann, H.: Chem. Ber. *112*, 1233 (1979)
73. Christl, M., Bruentrup, G.: ibid. *107*, 3908 (1974)
74. Christl, M. et al.: ibid. *110*, 3745 (1977)
75. Christl, M., Freitag, G., Bruentrup, G.: ibid. *111*, 2307 (1978)
76. Herbert, R., Christl, M.: ibid. *112*, 2012 (1979)
77. Christl, M., Freitag, G., Bruentrup, G.: ibid. *111*, 2320 (1978)
78. Christl, M., Leininger, H.: Tetrahedron Lett. *1979*, 1553
79. Leininger, H., Christl, M.: Angew. Chem. *92*, 466 (1980)

80. Christl, M., Brunn, E.: ibid. *93*, 474 (1981)
81. Gream, G. E., Smith, L. R., Meinwald, J.: J. Org. Chem. *39*, 3461 (1974)
82. Smith, L. R., Gream, G. E., Meinwald, J.: ibid. *42*, 927 (1977)
83. Hartke, K., Hansen, G., Kissel, T.: Liebigs Ann. Chem. *1980*, 1665
84. Roth, R. J., Katz, T. J.: J. Am. Chem. Soc. *94*, 4770 (1972)
85. Roth, R. J.: Synth. Commun. *9*, 751 (1979)
86. Roth, R. J., Woodside, A. B.: ibid. *10*, 645 (1980)
87. Katz, T. J., Nicolaou, K. C.: J. Am. Chem. Soc. *96*, 1948 (1974)
88. Christl, M., Lang, R., Herbert, R., Freitag, G.: Angew. Chem. *92*, 465 (1980)
89. Hogeveen, H., Kwant, P. W.: Tetrahedron Lett. *1973*, 3747
90. Hogeveen, H., Huurdeman, W. F. J.: J. Am. Chem. Soc. *100*, 860 (1978)
91. Heldeweg, R. F., Hogeveen, H., Schudde, E. P.: J. Org. Chem. *43*, 1912 (1978)
92. Heldeweg, R. F., Hogeveen, H.: ibid. *43*, 1916 (1978)
93. Giordano, C., Heldeweg, R. F., Hogeveen, H.: J. Am. Chem. Soc. *99*, 5181 (1977)
94. Kobayashi, Y. et al.: Tetrahedron Lett. *1975*, 3001
95. Masamune, S., Machiguchi, T., Aratani, M.: J. Am. Chem. Soc. *99*, 3524 (1977)
96. Kobayashi, Y. et al.: Tetrahedron Lett. *1976*, 2545
97. Warrener, R. N., Nun, E. E., Paddon-Row, M. N.: ibid. *1976*, 2639
98. Kobayashi, Y. et al.: ibid. *1977*, 1795
99. Kobayashi, Y. et al.: ibid. *1975*, 3819
100. Kobayashi, Y., Hanzawa, Y., Nakanishi, Y.: ibid. *1977*, 3571
101. Kobayashi, Y. et al.: ibid. *1976*, 2703
102. Kobayashi, Y., Kumadaki, I.: Advances in Heterocyclic Chem. *31*, 169 (1982)
103. Singh, B., Ullman, E. F.: J. Am. Chem. Soc. *89*, 6911 (1967)
104. van Tamelen, E. E., Whitesides, T. H.: ibid. *93*, 6129 (1971)
105. Chambers, R. D., Lindley, A. A., Fielding, H. C.: J. Fluorine Chem. *12*, 337 (1978)
106. Couture, A., Lablache-Combier, A.: J. Chem. Soc., Chem. Commun. *1971*, 891
107. Kellog, R. M. et al.: J. Org. Chem. *35*, 2737 (1970)
108. Kellog, R. M.: Tetrahedron Lett. *1972*, 1429
109. Wiebe, H. A., Braslavsky, S., Heicklen, J.: Can. J. Chem. *50*, 271 (1972)
110. Kobayashi, Y., et al.: Tetrahedron Lett. *1974*, 2841; Chem. Pharm. Bull. *23*, 2773 (1975)
111. Barltrop, J. A., Day, A. C., Irving, E.: J. Chem. Soc., Chem. Commun. *1979*, 881
112. Barltrop, J. A., Day, A. C., Irving, E.: ibid. *1979*, 966
113. Kobayashi, Y.: Tetrahedron Lett., in press
114. Hiraoka, H.: J. Chem. Soc., Chem. Commun. *1971*, 1610
115. Barltrop, J. A., Day, A. C., Ward, R. W.: ibid. *1978*, 131
116. Maeda, M., Kojima, M.: ibid. *1973*, 539; J. Chem. Soc. Perkin Trans. I *1977*, 239
117. Maeda, M., Kojima, M.: J. Chem. Soc., Perkin Trans. I *1978*, 685
118. Gotthardt, H., Reiter, F.: Chem. Ber. *112*, 1206 (1979)
119. Kobayashi, Y. et al.: J. Chem. Soc. Perkin Trans. I *1977*, 2355
120. Kobayashi, Y. et al.: J. Am. Chem. Soc. *99*, 7350 (1977)
121. Kobayashi, Y., Ando, A., Kumadaki, I.: J. Chem. Soc., Chem. Commun. *1978*, 509
122. Kobayashi, Y. et al.: J. Org. Chem. *45*, 2962 (1980)
123. Kobayashi, Y. et al.: ibid. *45*, 2966 (1980)
124. Kobayashi, Y. et al.: ibid. *45*, 2968 (1980)
125. Wirth, D., Lemal, D. M.: J. Am. Chem. Soc. *104*, 847 (1982)
126. Langanis, E. D., Lemal, D. M.: ibid. *102*, 6633, 6634 (1980)
127. Kobayashi, Y. et al.: J. Chem. Soc., Chem. Commun. *1981*, 1289
128. Kobayashi, Y. et al.: J. Am. Chem. Soc. *103*, 3958 (1981)
129. Kobayashi, Y. et al.: Tetrahedron Lett. *1975*, 1639
130. Ross, J. A., Seiders, R. P., Lemal, D. M.: J. Am. Chem. Soc. *98*, 4325 (1976)
131. Bushweller, C. H., Ross, J. A., Lemal, D. M.: *99*, 629 (1977)
132. Snyder, J. P., Halgren, T. A.: J. Am. Chem. Soc. *102*, 2861 (1980)
133. Corey, E. J., Streith, J.: ibid. *86*, 950 (1964)
134. Lin, C. Y., Kranz, A.: J. Chem. Soc., Chem. Commun. *1972*, 1111
135. Chapman, O. L., McIntosh, C. L., Pacansky, J.: J. Am. Chem. Soc. *95*, 614 (1973)
136. Pong, R. G. S. et al.: ibid. *99*, 4153 (1977)

137. Huang, B.-S. et al.: ibid. *99*, 4155 (1977)
138. Maier, G., Schafer, U.: Tetrahedron Lett. *1977*, 1053
139. Sharp, L. J., Hammond, G. S.: Mol. Photochem. *2*, 225 (1970)
140. Begley, W. J. et al.: J. Chem. Soc. Perkin Trans. I *1981*, 2620
141. Kaneko, C. et al.: J. Chem. Soc., Chem. Commun. *1980*, 1177
142. Hogeveen, H., Kok, D. M.: J. Org. Chem. *47*, 997 (1982)
143. Dennis, N., Katritzky, A. R., Wilde, H.: J. Chem. Soc. Perkin Trans. I *1976*, 2338
144. Beeken, P. et al.: J. Am. Chem. Soc. *101*, 6677 (1979)
145. Hasan, I., Fowler, F. W.: ibid. *100*, 6696 (1978)
146. Ogata, Y., Takagi, T.: J. Org. Chem. *43*, 944 (1978)
147. Warrener, R. N., Kretschmer, G., Paddon-Row, M. N.: J. Chem. Soc., Chem. Commun. *1977*, 806
148. Caplain, S., Lablache-Combier, A.: ibid. *1970*, 1247
149. Wilzbach, K. E., Rausch, D. J.: J. Am. Chem. Soc. *92*, 2178 (1970)
150. Dewar, M. J. S. et al.: J. Chem. Res. (S) *1978*, 26
151. Barlow, M. G., Dingwall, J. G., Haszeldine, R. N.: J. Chem. Soc., Chem. Commun. *1970*, 1580
152. Kobayashi, Y., Ohsawa, A., Iitaka, Y.: Tetrahedron Lett. *1973*, 2643
153. Kobayashi, Y. et al.: Chem. Pharm. Bull. *24*, 2219 (1976)
154. Kobayashi, Y., Ohsawa, A.: Chem. Pharm. Bull. *24*, 2225 (1976)
155. Chambers, R. D., Middleton, R., Corbally, R. P.: J. Chem. Soc., Chem. Commun. *1975*, 731
156. Chambers, R. D., Middleton, R.: ibid. *1977*, 154
157. Nishio, T. et al.: J. Chem. Soc. Perkin Trans. I *1980*, 607
158. Masamune, S., Nakamura, N., Sapadaro, J.: J. Am. Chem. Soc. *97*, 918 (1975)
159. Wildi, E. A., Carpenter, B. K.: Tetrahedron Lett. *1978*, 2469
160. Altman, L. J. et al.: J. Chem. Soc., Chem. Commun. *1968*, 686
161. Nunn, E. E., Warrener, R. N.: ibid. *1972*, 818
162. Allison, C. G. et al.: ibid. *1969*, 1200
163. Chambers, R. D., Musgrave, W. K. R., Srivastava, K. C.: ibid. *1971*, 264
164. Chambers, R. D., Maslakievicz, J. R., Srivastava, K. C.: J. Chem. Soc. Perkin Trans. I *1975*, 1130
165. Chambers, R. D., Maslakievicz, J. R.: J. Chem. Soc., Chem. Commun. *1976*, 1005
166. Chambers, R. D., Sargent, C. R.: J. Chem. Soc., Chem. Commun. *1979*, 446
167. Chambers, R. D., Musgrave, W. K. R., Sargent, C. R.: J. Chem. Soc. Perkin Trans. I *1981*, 1071
168. Chambers, R. D., Musgrave, W. K. R., Wood, D. E.: ibid. *1979*, 1978
169. Barlow, M. G., Haszeldine, R. N., Simon, C.: J. Chem. Soc. Perkin Trans. *1980*, 2254
170. Gompper, R., Schönafinger, K.: Chem. Ber. *112*, 1514 (1979)
171. Kobayashi, Y. et al.: Tetrahedron Lett. *1976*, 3915
172. Kobayashi, Y. et al.: J. Am. Soc. *99*, 8511 (1977)
173. Tinnemans, A. H. A., Neckers, D. C.: J. Org. Chem. *42*, 2374 (1977)
174. Davis, P. D., Neckers, D. C.: Tetrahedron Lett. *1978*, 2979
175. Reinhoudt, D. N., Kouwenhoven, C. G.: Tetrahedron *30*, 2431 (1974)
176. Reinhoudt, D. N., Kouwenhoven, C. G.: Tetrahedron Lett. *1974*, 2503
177. Kobayashi, Y., Hanzawa, Y.: ibid. *1978*, 4301
178. Uyegaki, M. et al.: ibid. *1976*, 4473
179. Murata, I., Nakasuji, K.: Top. Curr. Chem. *97*, 33 (1981)
180. Kobayashi, Y. et al.: J. Am. Chem. Soc. *101*, 6445 (1979)
181. Kobayashi, Y. et al.: Tetrahedron Lett. *1981*, 1113
182. Tsuchiya, T.: J. Soc. Org. Synth. Chem. Japan *39*, 99 (1981)
183. Tsuchiya, T. et al.: J. Chem. Soc., Chem. Commun. *1979*, 534
184. Tsuchiya, T., Kurita, J., Kojima, H.: ibid. *1980*, 444

Natural Vibrational Raman Optical Activity

Laurence D. Barron[1] and Julian Vrbancich[2]

1 Chemistry Department, The University, Glasgow G12 8QQ, U.K.
2 Research School of Chemistry, Australian National University, Canberra, A.C.T. 2600, Australia

Table of Contents

This article reviews all the published work concerned with the study of vibrational optical activity in chiral molecules from measurements of a small difference in the intensity of Raman scattering in right and left circularly polarized incident light. The history and basic theory are described briefly, followed by an account of the instrumentation and the precautions that must be observed in order to suppress spurious signals. The various theories that have been proposed in order to relate stereochemical features to the observations are then outlined, this being followed by a survey of all reported Raman optical activity spectra.

1 Introduction

Optical activity is an old subject dating back to the early years of the last century. But it is far from exhausted. Recent developments in optical and electronic technology have led to large increases in the sensitivity of conventional optical activity measurements, and have enabled completely new optical activity phenomena to be observed. Optical activity has been traditionally associated almost exclusively with electronic transitions, but one important advance over the past decade has been the extension of optical activity measurements into the vibrational spectrum using both infrared and Raman techniques. It is now apparent that the advent of vibrational optical activity has opened up a new world of fundamental studies and practical applications.

The Raman approach to vibrational optical activity is based on measurement of a small difference in the intensity of Raman-scattered light from chiral molecules in right and left circularly polarized incident light, and several reviews have appeared previously [1–5]. However, another review is now timely because important experimental and theoretical developments have since brought Raman optical activity (ROA) to a new level of maturity.

The significance of vibrational optical activity becomes apparent when it is compared with conventional electronic optical activity in the form of optical rotatory dispersion (ORD) and circular dichroism (CD) of visible and near-ultraviolet radiation. These conventional techniques have proved most valuable in stereochemistry, but since the electronic transition frequencies of most structural units in a molecule occur in inaccessible regions of the far-ultraviolet, they are restricted to probing chromophores and their immediate intramolecular environments. On the other hand, a vibrational spectrum contains bands from most parts of a molecule, so the measurement of vibrational optical activity should provide much more information.

The obvious method of measuring vibrational optical activity is by extending ORD and CD into the infrared. But in addition to the technical difficulties in manipulating polarized infrared radiation, there is a fundamental physical difficulty inherent in this approach: optical activity observables are proportional to the frequency of the *exciting* radiation, and infrared frequencies are orders of magnitude smaller than visible frequencies. This snag is circumvented using ROA because the Raman effect provides complete vibrational spectra using *visible* exciting light. Generally, the infrared and Raman techniques provide complementary approaches to natural vibrational optical activity. Thus at present, infrared CD has penetrated down to about 900 cm^{-1}, and although ROA covers the entire vibrational spectrum from about 50 to 4000 cm^{-1}, it is not as good as infrared CD for the region above 2000 cm^{-1} which contains fundamental CH and CD stretching bands. Infrared CD is also better for studying dilute non-aqueous solutions. On the other hand, ROA is at its best in the low-frequency region currently inaccessible to infrared CD where skeletal deformation and torsion modes, which carry stereochemical information most directly, occur. There is also the possibility of applying ROA to biological molecules in aqueous media, not only through transparent scattering, but also through resonance scattering, which provides the exciting possibility of directly probing the stereochemistry at sites of biological function. It should be mentioned that, as infrared

CD measurements penetrate to lower frequencies, the fundamental difficulty mentioned above does not appear to be a serious limitation since large effects are still observed: this might be due to the fact that deformation coordinates, which can generate much larger vibrational optical activity than stretch coordinates, start to make significant contributions to the normal modes. Several recent reviews discuss infrared CD and ROA from a unified standpoint [5-8].

Rayleigh and Raman optical activity has an interesting and chequered history. In the late 19th century, Lord Rayleigh showed that the light scattered at right angles from spherical molecules should be completely linearly polarized perpendicular to the scattering plane. The observed imperfection in this linear polarization in light scattered from gases was ascribed to a lack of spherical symmetry in the optical properties of the molecules. In 1923 Gans [9] discussed the possibility of additional polarization phenomena in light scattered from optically active molecules: in particular, he investigated the contribution to Rayleigh scattering from the optical activity tensor alone and claimed to have observed its effect on the depolarization ratio; but de Malleman [10] pointed out that the anomalies originated in optical rotation of the incident and scattered beams. Also, Gans omitted terms describing interference between the polarizability and optical activity tensors which are central to the discussion of Rayleigh and Raman optical activity. Shortly after the discovery of the Raman effect, Bhagavantam and Venkateswaran [11] found differences in the relative intensities of some of the vibrational Raman lines of two enantiomers, but these were attributed subsequently to impurities. Although he had no theory, Kastler [12] though that a difference might exist between the Raman spectra of optically active molecules in right and left circularly polarized incident light, and attempted unsuccessfully to observe it. Perrin [13] alluded to the existence of additional polarization effects in light scattered from optically active molecules, but it was Atkins and Barron [14] who showed explicitly that interference between the molecular polarizability and optical activity tensors leads to a dependence of the scattered intensity on the degree of circular polarization of the incident light and to a circularly polarized component in the scattered light. Subsequently, Barron and Buckingham [15] developed the theory of the Rayleigh and Raman circular intensity difference (CID) defined by

$$\Delta_\alpha = (I_\alpha^R - I_\alpha^L)/(I_\alpha^R + I_\alpha^L) \tag{1.1}$$

where I_α^R and I_α^L are the scattered intensities with α-polarization in right and left circularly polarized incident light. The first reported Raman CID spectra by Bosnich et al. [16] and Diem et al. [17] originated in experimental artefacts, but the spectra reported shortly afterwards by Barron et al. [18-20] in simple chiral molecules such as α-phenylethylamine and α-pinene have been confirmed as the first genuine observations [21-24].

Since all molecules can show ORD and CD in a magnetic field (the Faraday effect), it is not surprising that all molecules in a strong magnetic field parallel to the incident light beam should show Rayleigh and Raman optical activity. For further details we refer to a recent article that reviews magnetic Raman optical activity and other field-induced optical activity phenomena in light scattering [25].

2 Basic Theoretical Results

We use a semi-classical formulation in which the origin of scattered light is considered to be the characteristic radiation fields generated by the oscillating electric and magnetic multipole moments induced in a molecule by the electromagnetic fields of the incident light wave. The induced multipole moments are related to the electric and magnetic vectors of the incident light wave by property tensors. We require in particular the polarizability tensor $\alpha_{\alpha\beta}$ and the optical activity tensors $G'_{\alpha\beta}$ and $A_{\alpha\beta\gamma}$, for which time-dependent perturbation theory provides the following quantum-mechanical expressions [26,5)]

$$\alpha_{\alpha\beta} = \frac{2}{\hbar} \sum_{j \neq n} \frac{\omega_{jn}}{\omega_{jn}^2 - \omega^2} \operatorname{Re}(\langle n|\mu_\alpha|j\rangle\langle j|\mu_\beta|n\rangle) \tag{2.1 a}$$

$$G'_{\alpha\beta} = -\frac{2}{\hbar} \sum_{j \neq n} \frac{\omega}{\omega_{jn}^2 - \omega^2} \operatorname{Im}(\langle n|\mu_\alpha|j\rangle\langle j|m_\beta|n\rangle) \tag{2.1 b}$$

$$A_{\alpha\beta\gamma} = \frac{2}{\hbar} \sum_{j \neq n} \frac{\omega_{jn}}{\omega_{jn}^2 - \omega^2} \operatorname{Re}(\langle n|\mu_\alpha|j\rangle\langle j|\theta_{\beta\gamma}|n\rangle) \tag{2.1 c}$$

where ω is the angular frequency, $|n\rangle$ and $|j\rangle$ are the ground and excited molecular states, and μ_α, m_α and $\theta_{\alpha\beta}$ are the electric dipole, magnetic dipole and electric quadrupole moment operators defined by

$$\mu_\alpha = \sum_i e_i r_{i\alpha} \tag{2.2 a}$$

$$m_\alpha = \sum_i \frac{e_i}{2m_i} \varepsilon_{\alpha\beta\gamma} r_{i\beta} p_{i\gamma} \tag{2.2 b}$$

$$\theta_{\alpha\beta} = \frac{1}{2} \sum_i e_i (3r_{i\alpha} r_{i\beta} - r_i^2 \delta_{\alpha\beta}) \tag{2.2 c}$$

e_i, m_i, $r_{i\alpha}$ and $p_{i\alpha}$ are the charge, mass, position vector and linear momentum of the ith charge. We are using a cartesian tensor notation in which $\delta_{\alpha\beta}$ is the unit second

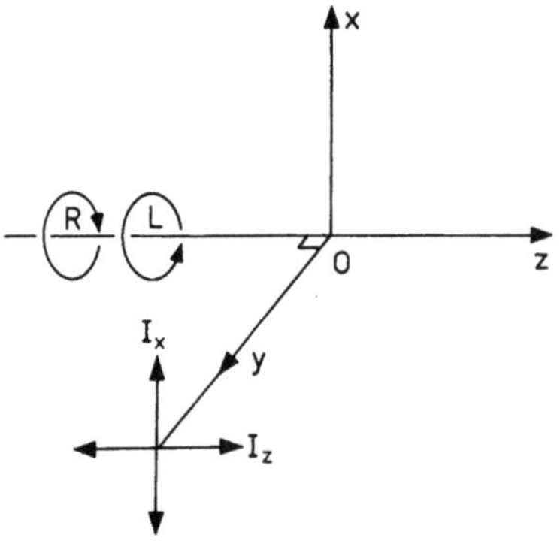

Fig. 1. The geometry for circular differential scattering at 90°

rank symmetric tensor and $\varepsilon_{\alpha\beta\gamma}$ is the unit third rank antisymmetric tensor, and a repeated Greek suffix denotes summation over the three components [5].

By considering the dependence of the scattered intensity on the polarization state of the incident light beam, the components of the CID (1.1) for light scattered at 90° and polarized perpendicular (Δ_x) and parallel (Δ_z) to the scattering plane yz (Fig. 1) are found to be [15,1-3,5]

$$\Delta_x(90°) = \frac{2\left(7\alpha_{\alpha\beta}G'_{\alpha\beta} + \alpha_{\alpha\alpha}G'_{\beta\beta} + \frac{1}{3}\omega\alpha_{\alpha\beta}\varepsilon_{\alpha\gamma\delta}A_{\gamma\delta\beta}\right)}{c(7\alpha_{\lambda\mu}\alpha_{\lambda\mu} + \alpha_{\lambda\lambda}\alpha_{\mu\mu})} \quad (2.3\text{ a})$$

$$\Delta_z(90°) = \frac{4\left(3\alpha_{\alpha\beta}G'_{\alpha\beta} - \alpha_{\alpha\alpha}G'_{\beta\beta} - \frac{1}{3}\omega\alpha_{\alpha\beta}\varepsilon_{\alpha\gamma\delta}A_{\gamma\delta\beta}\right)}{2c(3\alpha_{\lambda\mu}\alpha_{\lambda\mu} - \alpha_{\lambda\lambda}\alpha_{\mu\mu})} \quad (2.3\text{ b})$$

It is useful to rewrite the CIDs (2.3) in terms of the following tensor invariants with respect to molecule-fixed axes X, Y, Z [5,27]

$$\alpha = \frac{1}{3}\alpha_{\alpha\alpha} = \frac{1}{3}(\alpha_{XX} + \alpha_{YY} + \alpha_{ZZ}) \quad (2.4\text{a})$$

$$\begin{aligned}\beta(\alpha)^2 &= \frac{1}{2}(3\alpha_{\alpha\beta}\alpha_{\alpha\beta} - \alpha_{\alpha\alpha}\alpha_{\beta\beta}) \\ &= \frac{1}{2}[(\alpha_{XX} - \alpha_{YY})^2 + (\alpha_{XX} - \alpha_{ZZ})^2 + (\alpha_{YY} - \alpha_{ZZ})^2 \\ &\quad + 6(\alpha_{XY}^2 + \alpha_{XZ}^2 + \alpha_{YZ}^2)]\end{aligned} \quad (2.4\text{b})$$

$$G' = \frac{1}{3}G'_{\alpha\alpha} = \frac{1}{3}(G'_{XX} + G'_{YY} + G'_{ZZ}) \quad (2.4\text{c})$$

$$\begin{aligned}\beta(G')^2 &= \frac{1}{2}(3\alpha_{\alpha\beta}G'_{\alpha\beta} - \alpha_{\alpha\alpha}G'_{\beta\beta}) \\ &= \frac{1}{2}[(\alpha_{XX} - \alpha_{YY})(G'_{XX} - G'_{YY}) + (\alpha_{XX} - \alpha_{ZZ})(G'_{XX} - G'_{ZZ}) \\ &\quad + (\alpha_{YY} - \alpha_{ZZ})(G'_{YY} - G'_{ZZ}) + 6(\alpha_{XY}G'_{XY} + \alpha_{XZ}G'_{XZ} \\ &\quad + \alpha_{YZ}G'_{YZ})]\end{aligned} \quad (2.4\text{d})$$

$$\begin{aligned}\beta(A)^2 &= \frac{1}{2}\omega\alpha_{\alpha\beta}\varepsilon_{\alpha\gamma\delta}A_{\gamma\delta\beta} \\ &= \frac{1}{2}\omega[(\alpha_{YY} - \alpha_{XX})A_{ZXY} + (\alpha_{XX} - \alpha_{ZZ})A_{YZX} \\ &\quad + (\alpha_{ZZ} - \alpha_{YY})A_{XYZ}]\end{aligned} \quad (2.4\text{e})$$

Since $\varepsilon_{\alpha\gamma\delta}A_{\gamma\delta\beta}$ is traceless, there is no corresponding isotropic tensor analogous to α and G'. In a principal axis system that diagonalizes $\alpha_{\alpha\beta}$, the terms in (2.4) involving off-diagonal components of $\alpha_{\alpha\beta}$ vanish. The CIDs (2.3) now become [5,27]

$$\Delta_x(90°) = \frac{2[45\alpha G' + 7\beta(G')^2 + \beta(A)^2]}{c[45\alpha^2 + 7\beta(\alpha)^2]} \tag{2.5a}$$

$$\Delta_z(90°) = \frac{12\left[\beta(G')^2 - \frac{1}{3}\beta(A)^2\right]}{6c\beta(\alpha)^2} \tag{2.5b}$$

It is also unseful to have the CIDs for forward and backward scattering [5]

$$\Delta(0°) = \frac{4[45\alpha G' + \beta(G')^2 - \beta(A)^2]}{c[45\alpha^2 + 7\beta(\alpha)^2]} \tag{2.5c}$$

$$\Delta(180°) = \frac{24\left[\beta(G')^2 + \frac{1}{3}\beta(A)^2\right]}{c[45\alpha^2 + 7\beta(\alpha)^2]} \tag{2.5d}$$

We refer to Andrews [28] for further discussion of the dependence of the CID components on the scattering angle, and the extraction of the tensor invariants. Also, the CIDs in forward and backward scattering have been discussed in the context of coherent antistokes Raman scattering [29] and the Raman-induced Kerr effect [30], respectively.

As written, the CIDs (2.3) and (2.5) apply to *Rayleigh* scattering. The same expression can be used for *Raman* optical activity if the property tensors are replaced by corresponding vibrational Raman transition tensors. This enables us to deduce the basic symmetry requirements for natural vibrational ROA [15,5]: the same components of $\alpha_{\alpha\beta}$ and $G'_{\alpha\beta}$ must span the irreducible representation of the particular normal coordinate of vibration. This can only happen in the chiral point groups C_n, D_n, O, T, I (which lack improper rotation elements) in which polar and axial tensors of the same rank, such as $\alpha_{\alpha\beta}$ and $G'_{\alpha\beta}$ (or $\varepsilon_{\alpha\gamma\delta}A_{\gamma\delta\beta}$) have identical transformation properties. Thus, all the Raman-active vibrations in a chiral molecule should show Raman optical activity.

There is, in fact, an alternative to the CID for obtaining ROA spectra: measurement of the degree of circularity (the intensity of the circularly polarized component) of the scattered beam [31,5,27]. The degree of circularity is specified by the ratio S_3/S_0, where S_3 is the Stokes parameter that corresponds to the excess in intensity transmitted by a device which accepts right circularly polarized light over that transmitted by a device which accepts left, and S_0 is the Stokes parameter that corresponds to the total intensity. Thus for forward and backward scattering, S_3/S_0 is precisely equal to (2.5c) and (2.5d), respectively; and for 90° scattering S_3/S_0 is equal to (2.5a) in incident light linearly polarized perpendicular to the scattering plane, and is equal to (2.5b) in incident light linearly polarized parallel to the scattering plane [31,5].

A central feature in the development of theories of the generation of ROA within chiral molecules has been the realization that the optical activity tensors are origin-dependent. If the origin is moved from **O** to a point **O** + **a**, where **a** is some constant vector, it is found that [32,5]

$$\alpha_{\alpha\beta} \rightarrow \alpha_{\alpha\beta} \quad (2.6a)$$

$$G'_{\alpha\beta} \rightarrow G'_{\alpha\beta} + \frac{1}{2}\omega\varepsilon_{\beta\gamma\delta}a_{\gamma}\alpha_{\alpha\delta} \quad (2.6b)$$

$$A_{\alpha\beta\gamma} \rightarrow A_{\alpha\beta\gamma} - \frac{3}{2}a_{\beta}\alpha_{\alpha\gamma} - \frac{3}{2}a_{\gamma}\alpha_{\alpha\beta} + a_{\delta}\alpha_{\alpha\delta}\delta_{\beta} \quad (2.6c)$$

3 Instrumentation

All the ROA spectra published to date have been obtained by measuring the circular intensity difference. No reports have appeared of the alternative method involving measurement of the circularly polarized component of the scattered light, presumably on account of the difficulty in making accurate polarization measurements on divergent scattered light. This alternative might be advantageous in certain one-off experiments, such as ROA studies of gases, since a multiple pass cell could then be used; but for most routine studies measurement of the CID is likely to remain the preferred technique.

The basic Raman CID experiment is very simple, at least in principle. In the first generation of instruments, an electro-optic modulator, for switching the polarization of the laser beam between right and left circular, was added to a conventional scanning Raman spectrometer arranged to detect scattered light at 90° and equipped for photon counting. Using synchronous detection techniques, the difference in the Raman-scattered intensity in right and left circularly polarized incident light is obtained directly. The first successful observations of Raman CID in Cambridge [18] and Berkeley [21] were made with instruments of this type, and similar instruments have been used in Toledo [22], Toronto [23], Tokyo [24] and Syracuse [33]. In the second generation of instruments, developed in Toronto [23,34] and Fribourg [35], the scanning spectrometer was replaced by an optical multichannel spectrograph which can provide and increase in speed of one or two orders of magnitude. Both types of instrument are described in more detail below.

Raman optical activity has only been measured so far in pure liquids and strong solutions. Crystals and powders are harder to study: crystals must be polished and oriented carefully to eliminate artefacts, whereas multiple scattering in powders depolarizes the incident light. It would be of great interest to measure pure rotational, and rotational-vibrational, ROA in gases, but insufficient scattered intensity has so far prevented this. An additional complication in *resonance* scattering is that circular dichroism of the incident beam can contribute to the measured circular intensity difference.

3.1 Scanning ROA Methods

After being closely involved with the first successful ROA experiments in Cambridge, one of the authors (L.D.B.) moved to Glasgow in 1975. Since the technology of multichannel light detectors was still in a state of flux at that time, he decided not to attempt to develop a multichannel ROA instrument, but instead to extend, using the principles of 'brute force' and simplicity, the scanning type of instrument that had already proved successful. The soundness of that decision is born out by the fact that most of the ROA spectra published to date have been generated since 1975 by the Glasgow scanning instrument.

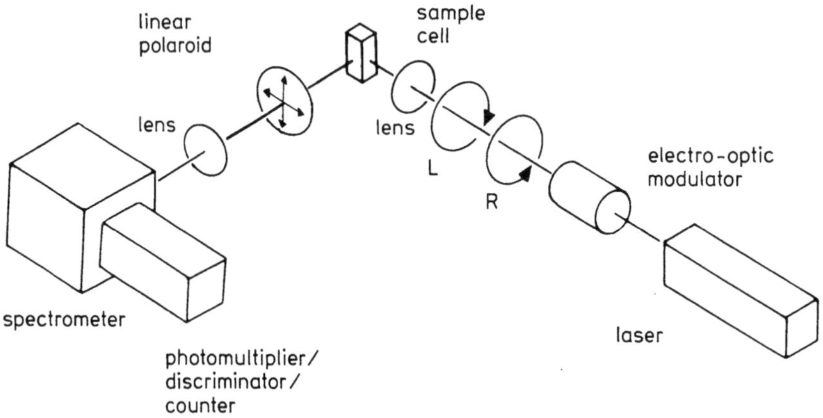

Fig. 2. Basic layout of the scanning Raman CID spectrometer

The basic layout of the Glasgow scanning instrument is shown in Fig. 2. The linearly polarized output from an argon-ion laser (Spectra-Physics Model 171-08) passes through an electro-optic modulator (Electro-Optic Develoments Model PC 105) which has a longitudinal electric field generated by gold ring electrodes on the end faces of a KDP crystal, and is driven by a high voltage linear differential amplifier (Electro-Optic Developments Model LDA 10). This amplifier has adjustable positive and negative limits and generates an ultra-stable high voltage square wave which enables the incident linearly polarized laser beam to be switched between pure right and pure left circular polarization at a suitable frequency (we use about 100 Hz). The laser is focused through the sample, which is usually held in a fluorescence cell, and the light scattered at 90° passes through a linear polaroid disc (HN 38) and is collected by a triple monochromator spectrometer (Coderg Model T800). The analysed light is detected by a photomultiplier tube (EMI 9658A) specially selected for high quantum efficiency in the blue-green region. The voltage pulses from individual photon strikes are passed through a discriminator (SSR Model 1120) to suppress background noise pulses, and the pulses originating from left and right circularly polarized incident radiation are counted is separate channels by synchronizing a dual-channel counter (SSR Model 1110) with the low-voltage input to the electro-optic modulator drive amplifier. After accumulating a statistically-significant number of counts at a given spectral point, the sum of the two counters $I^R + I^L$ and the difference $I^R - I^L$ are read. The statistical significance of N counts is determined by the standard deviation $N^{1/2}$. Thus on 10^8 counts, the 'statistical noise' is $N^{1/2}/N = 10^{-4}$, an order of magnitude smaller than the larger CIDs expected. On typical Raman bands at an effective resolution of $10\ cm^{-1}$, 10^8 counts can be collected in about one minute using 3W laser power at 488 nm, and well-defined $I^R - I^L$ spectra can be recorded automatically at a scan rate of $1\ cm^{-1} \cdot min^{-1}$.

3.2 Optical Multichannel Methods

An optical multichannel Raman instrument functions as a spectrograph rather than as a monochromator due to the absence of an exit slit. The Raman light at the output of a grating instrument is dispersed across a detector consisting of an electronic image sensor that functions as an 'electronic photographic plate'. For optical activity measurements, the components between the laser and the sample are essentially the same as described in the previous section for the scanning instrument.

The first multichannel ROA measurements to be reported were obtained in the Toronto laboratory of Moskovits et al. [23,34]. Their detector was an intensified vidicon tube which was placed at the output port of the first monochromator of a double-monochromator Raman spectrometer. Vidicon detectors are made up of many discrete photodetectors which combine the functions of radiation sensor and charge storage: light falling on the photodetectors discharges them, and they are then recharged using a scanning electron beam with the charging current required for each resolution element being proportional to the light intensity previously incident on that element [36]. Despite certain problems intrinsic to vidicons (such as 'lag' and 'blooming'), Moskovits et al. obtained good ROA spectra by collecting data simultaneously over a $350\ cm^{-1}$ portion of the spectrum, and estimated that they had achieved a roughly tenfold increase in speed over their earlier scanning instrument.

The *self-scanning* solid-state silicon photodiode array is a more recent type of electronic image sensor that has considerable advantages over the vidicon tube [37]. It combines the third basic function of readout with the other two (radiation sensor and charge storage) in a single integrated circuit. By using this type of detector coupled to an image intensifier, Hug and Surbeck [35] achieved a roughly 200-fold increase in speed over scanning ROA instruments, which enabled them to record spectra covering $2100\ cm^{-1}$ in less than two hours using only 50 mW of laser power. This work was a tour de force since, as discussed below, they were also able to achieve an almost total suppression of artefacts. More details of the Fribourg instrument have recently been reported [38], together with an account of a version that has successfully measured ROA in backscattering. Unfortunately, these instruments have now been destroyed by fire [38]!

3.3 Experimental Quantities

There are currently two main conventions for displaying ROA spectra. The authors take the dimensionless CID (1.1) as the experimentally-relevant quantity, but present raw photon count spectra of $I_\alpha^R + I_\alpha^L$ and $I_\alpha^R - I_\alpha^L$ (from our scanning instrument) separately because the background must be subtracted from $I_\alpha^R + I_\alpha^L$ before calculating Δ_α. Also, we present the $I_\alpha^R + I_\alpha^L$ spectra on a linear scale, whereas the $I_\alpha^R - I_\alpha^L$ spectra are presented on a scale that is linear within each decade range but logarithmic between ranges. This linear-logarithmic scale enables the exponent in the $I_\alpha^R - I_\alpha^L$ photon count to be recorded, but has the disadvantage that the spectrum takes on a 'stretched out' appearance in which the statistical noise level looks much greater than it really is since, on all but the weakest bands, only values within the $\pm 10^4$ ranges are significant. However, this presentation has the distinct advantage that all of the data is provided in a readily discernable form: the large dynamic range of Raman spectra means that, on very weak Raman bands, significant $I^R - I^L$ features can be lost in the baseline of a linear display, but are amplified on the inner range of the linear-logarithmic display.

Hug and Surbeck [35] have proposed the use of $\Delta d\sigma = d\sigma^L - d\sigma^R$, the difference of the Raman differential scattering cross sections in left and right circularly polarized incident light. This is the Raman equivalent of the circular dichroism $\Delta\varepsilon = \varepsilon^L - \varepsilon^R$, where ε is the decadic molar extinction coefficient and, unlike the *measured* (but not the theoretical) $I^R - I^L$ which depends on both sample and instrumental factors, is solely a molecular parameter. They introduced a chirality number q defined by

$$q = \frac{\Delta\, d\sigma}{d\sigma} = \frac{d\sigma^L - d\sigma^R}{\frac{1}{2}(d\sigma^L + d\sigma^R)} \tag{3.1}$$

which has an advantage over the measured CID Δ in that it represents an enantiomerically pure sample, for which $q = -2\Delta$.

However, both the sign convention for q and the factor of $^1/_2$ in the denominator

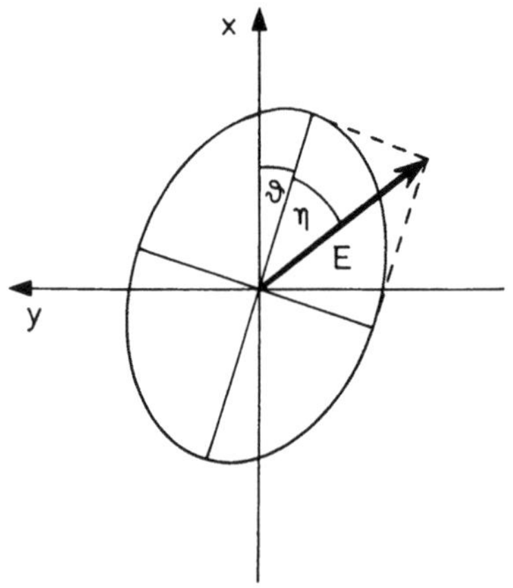

Fig. 3. The polarization ellipse referred to space-fixed axes x and y. The propagation direction z is out of the plane of the paper. η is the ellipticity and θ is the azimuth

are moot points [39], the main criticism being that they lead to difficulties with the principal of reciprocity when the alternative Raman optical activity experiment is considered, namely measurement of the circularly polarized component of the scattered light [31,27,5].

3.4 Artefacts

From the first, ROA has been plagued by artefacts. The first genuine observations were only achieved after the origin of the major sources of artefacts were understood in terms of the polarization-dependence of the isotropic and anisotropic contributions to scattering at 90° [40,18]. The intensity components of the light scattered at 90° linearly polarized perpendicular (I_x) and parallel (I_z) to the scattering plane yz are [40,18,1]

$$I_x = \frac{1}{2}\left(\frac{\varepsilon\varepsilon_0}{\mu\mu_0}\right)^{1/2}\frac{1}{30}\left(\frac{\omega^2\mu_0 E^{(0)}}{4\pi R}\right)^2\left\{15\alpha^2 + \frac{7}{3}\beta(\alpha)^2 + \left[15\alpha^2 + \frac{1}{3}\beta(\alpha)^2\right]P\cos 2\eta\cos 2\theta + \frac{2}{c}\left[\frac{7}{3}\beta(G')^2 + 15\alpha G' + \frac{1}{3}\beta(A)^2\right]P\sin 2\eta\right\} \quad (3.2\,a)$$

$$I_z = \left(\frac{\varepsilon\varepsilon_0}{\mu\mu_0}\right)^{1/2}\frac{1}{30}\left(\frac{\omega^2\mu_0 E^{(0)}}{4\pi R}\right)^2\left\{\beta(\alpha)^2 + \frac{2}{c}\left[\beta(G')^2 - \frac{1}{3}\beta(A)^2\right]P\sin 2\eta\right\} \quad (3.2\,b)$$

where P is the degree of polarization of the incident beam (P = 1 for a pure polarized beam and 0 for an unpolarized beam) and θ and η are the azimuth and ellipticity of the incident polarization ellipse defined in Fig. 3. Thus isotropic scattering through α^2 makes the largest contribution to the polarized intensity I_x, whereas anisotropic scattering through $\beta(\alpha)^2$ makes the largest contribution to the depolarized intensity I_z. The optically active scattering arises from $\alpha G'$, $\beta(G')^2$ and $\beta(A)^2$, and since these terms depend on $\sin 2\eta$ the sign of their contribution to the scattered intensity is different in right and left circularly polarized light, giving rise to the CID. Imperfections in the modulation system and the optical train can generate large artefacts in the modulated part of I_x through the second term in (3.2a) since this contains contributions from intense α^2 and $\beta(\alpha)^2$ scattering which depend on the polarization properties of the light beam in the form $P\cos 2\eta\cos 2\theta$. The modulated part of I_x is much less susceptible to artefacts. Perfect square wave modulation can eliminate the artefacts from the term in $P\cos 2\eta\cos 2\theta$ since $\cos 2\eta$ is then always zero. Also the dependence on $\cos 2\theta$ indicates that artefacts can be further suppressed by inclining the plane of polarization of the laser beam at 45° to the scattering plane before it enters the modulator. The secret of success in the original Cambridge experiments was therefore the use of square-wave modulation to switch the polarization of the laser beam between right and left circular polarization with gating of the photon counters to avoid collecting counts at the crossover points.

The extent of the artefact problem depends on the polarization characteristics of the particular Raman band. If the band is strongly polarized, isotropic scattering through α^2 contributes most intensity so that I_x is determined mainly by $\alpha^2(1 + P \cos 2\eta \cos 2\theta)$ and Δ_x is very susceptible to artefacts. Δ_z is also susceptible since part of I_z in strongly polarized bands originates in leakage of I_x through the analyzer. If the band is weakly polarized or depolarized, anisotropic scattering through $\beta(\alpha)^2$ contributes most intensity so that I_x is determined mainly by $\beta(\alpha)^2$ $(7 + P \cos 2\eta \cos 2\theta)$ and Δ_x is consequently less susceptible to artefacts: Δ_z is now even less susceptible, and artefacts can easily be held to $\pm 1 \times 10^{-4}$.

Because of these problems, only depolarized CIDs have been measured routinely on the first generation of instruments. This is not too restrictive, however, since depolarized CIDs are usually easier to interpret and no stereochemical information is lost; but for fundamental studies it is necessary to have polarized CIDs as well.

Hug extended the considerations outlined above to include in detail the effect of the finite solid angle in the cone of scattered light that enters the collection optic [41]. His analysis was based on a treatment of the angular distribution of Raman intensity in terms of differential scattering cross-sections, which is a little different from the treatment used here. However, the same results can be obtained by generalizing the present approach [42]. The Stokes parameters are calculated for light scattered into an arbitrary direction specified by the polar and azimuthal angles ξ and α relative to the space-fixed axes x, y, z with the azimuth of the polarization ellipse of the incident light beam still defined relative to the x axis (Fig. 4). The intensity components $I_x^d(\xi, \alpha, R)$ and $I_z^d(\xi, \alpha, R)$ are then calculated, and the quantity of energy dW passing in unit time across a surface element $R^2\, d\Omega$ associated with a cone of solid

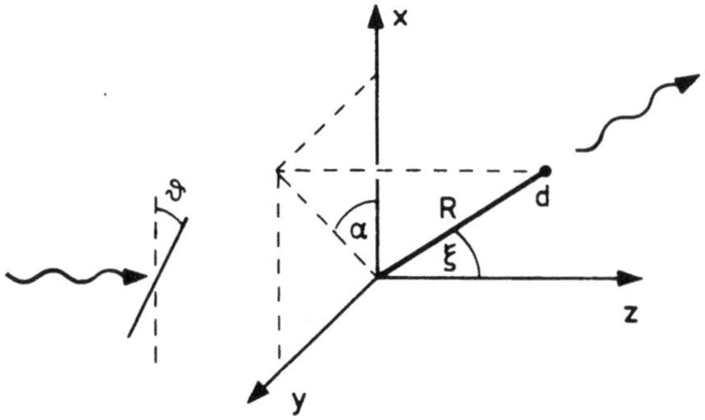

Fig. 4. An arbitrary scattering direction specified by polar and azimuthal angles ξ and α

angle $d\Omega = \sin \xi\, d\xi\, d\alpha$ about the scattering direction is obtained by multiplying the intensity by $R^2\, d\Omega$. For the case of 90° scattering along the y axis (Fig. 5), the total radiation energy per unit time entering the collection optic is obtained by integrating dW between the limits $\xi_0 + \Delta\xi/2$ and $\xi_0 - \Delta\xi/2$, and $\alpha_0 + \Delta\alpha/2$ and $\alpha_0 - \Delta\alpha/2$, where ξ_0, α_0 specify the axis of the collection optic and $\Delta\xi$, $\Delta\alpha$ its finite cone of collection. If P_R, η_R, θ_R and P_L, η_L, θ_L are the polarization parameters of the

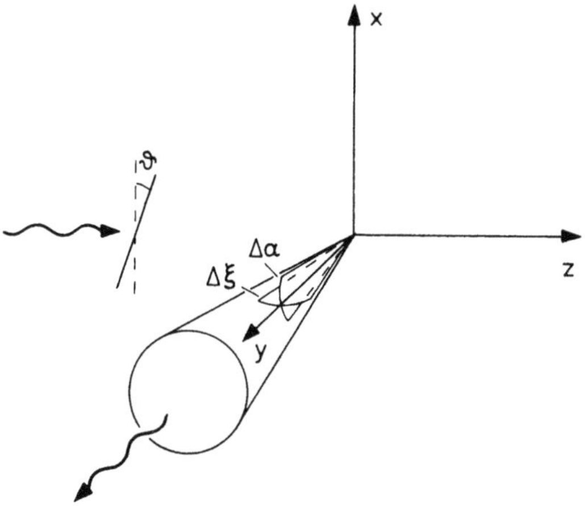

Fig. 5. The cone of scattered light collected around the y axis ($\xi_0 = \alpha_0 = 90°$)

incident beam in the 'right' and 'left' phases, the *spurious* differences in the radiation energy collected per unit time through an analyzer with its axis perpendicular (W_x) and parallel (W_z) to the scattering plane yz are found to be[42)]

$$W_x^R - W_x^L = \left(\frac{\varepsilon\varepsilon_0}{\mu\mu_0}\right)^{1/2} \frac{1}{30} \left(\frac{\omega^2\mu_0 E^{(0)}}{4\pi}\right)^2$$
$$\times\ \beta^2\left(\frac{1}{\varrho} - 1\right) (P_R \cos 2\eta_R \cos 2\theta_R - P_L \cos 2\eta_L \cos 2\theta_L)$$
$$\times\ \sin \Delta\alpha \sin (\Delta\xi/2) \qquad (3.3\,a)$$

$$W_z^R - W_z^L = -\frac{1}{3}\left(\frac{\varepsilon\varepsilon_0}{\mu\mu_0}\right)^{1/2} \frac{1}{30} \left(\frac{\omega^2\mu_0 E^{(0)}}{4\pi}\right)^2$$
$$\times\ \beta^2\left(\frac{1}{\varrho} - 1\right) (P_R \cos 2\eta_R \cos 2\theta_R - P_L \cos 2\eta_L \cos 2\theta_L)$$
$$\times\ \sin \Delta\alpha \sin^3 (\Delta\xi/2) \qquad (3.3b)$$

where ϱ is the depolarization ratio in incident light linearly polarized perpendicular to the scattering plane. These equations are equivalent to those given by Hug[41)]. Of course, if it were possible to switch between pure right and pure left incident circular polarization all the terms in (3.3) would vanish independently because $\eta = \pm 45°$.

It seems likely that the dominant artefacts are produced by linearly polarized contaminants in the incident circularly polarized beam, with different azimuths in the right and left phases. Hug noticed that, if the intensities of the linearly polarized contaminants in the two phases are identical, a twin-lens collection system comprising two identical collection optics with their axes in the xy plane and inclined at $\pm 45°$ to the y axis will eliminate the spurious intensity difference. For the collection optic with its axis at $+45°$ to the y axis, the incident azimuth is effectively $\theta - 45°$, whereas for the optic with its axis at $-45°$ the azimuth is effectively $\theta + 45°$, so that the spurious intensity difference is proportional to

$$[\cos 2(\theta_R - 45°) - \cos 2(\theta_L - 45°)] + [\cos 2(\theta_R + 45°) - \cos 2(\theta_L + 45°)] = 0 \quad (3.4)$$

for all values of θ_R and θ_L. This particular device could, of course, have been deduced from the earlier result 3.2 since this also displays the $\cos 2\theta$ dependence. The new feature of 3.3 is the explicit dependence of the artefacts on the angles $\Delta\alpha$ and $\Delta\xi$ defining the cone of collected light. The dependence on $\sin \Delta\alpha$ indicates that a uniform collection in a semicircle ($\Delta\alpha = 180°$) or a complete circle ($\Delta\alpha = 360°$) in the xy plane will eliminate these artefacts. The dependence on $\sin(\Delta\xi/2)$ for the polarized artefact and on $\sin^3(\Delta\xi/2)$ for the depolarized artefact indicates that these artefacts will be suppressed by keeping the collection angle in the scattering plane as small as possible, the depolarized artefact being suppressed most. Hug used these considerations, and others, to construct an ROA instrument in which both depolarized and polarized artefacts were almost completely suppressed, and demonstrated convincingly the value of a double-lens collection optic [41]. The semicircular light collection strategy is more difficult to put into practice, and has not yet been demonstrated.

4 Generation within Chiral Molecules

Theories of the generation of vibrational optical activity within chiral molecules are still in a state of flux. There are two main reasons for this. First, the theories are difficult, involving as they do the theories of both the electronic and vibrational states of low symmetry molecules. Second, there is not yet a sufficiently large body of experimental data against which the theories can be exhaustively tested. Nonetheless, significant progress has been made, and we review here the several different approaches that have been proposed for the Raman case. A much more detailed review of theories of vibrational optical activity, embracing infrared CD as well as ROA, has recently been published by Polavarapu [8].

In a completely asymmetric molecule (C_1), every internal coordinate contributes in some measure to each normal coordinate of vibration, which means that simple models based on local group frequency approximations do not often provide definitive interpretations of observed features. Simple models have, however, been important in establishing the physical basis of general theories, based on a full normal coordinate analysis, that can compute the sign and magnitude of the vibrational optical activity in every normal mode.

In the Raman case, three distinct general computational theories have been proposed: the bond polarizability theory, the atom dipole interaction theory and localized molecular orbital theories. In the first and third of these the normal modes of vibration, and hence the vibrational quantum states, must embrace a chiral nuclear framework. They are therefore analogous to the 'inherently chiral chromophore' model of electronic optical activity in which the electronic states are delocalized over a chiral nuclear framework [5]. No account is taken of electronic chirality in that the molecule is broken down into locally achiral atoms, bonds or groups. Thus the first and third computational theories are based on an extreme description which invokes the modulation of an achiral electronic distribution by a chiral normal mode of vibration. Another extreme description can be envisaged in which electronic chirality, generated through any of the standard models of electronic optical activity (electronic delocalization over a chiral nuclear framework, or non-bonded static and dynamic coupling between achiral groups) is modulated by a normal mode that is dominated by an internal coordinate localized on an achiral part of the framework. In practice it may not be possible to isolate these two extreme situations, but one or other may dominate in individual instances. At the time of writing the vibrational chirality description appears to provide the best description of most observed vibrational optical activity features. As discussed below, the second computational theory involves both electronic chirality (in the form of dynamic coupling between atoms) and vibrational chirality.

4.1 The Bond Polarizability Theory

The bond polarizability theory has developed out of a synthesis of two, at first sight distinct, models of ROA: the *two-group* model [43] and the *inertial* model [44]. In the former, Rayleigh and Raman optical activity originates in interference between waves scattered independently from two chiral anisotropic groups which together constitute a chiral structure: but unlike natural electronic optical rotation or circular dichroism, no coupling is required between the two groups [5]. In the latter, it is generated by the changing interaction of the radiation field with a chiral molecular framework as the framework twists in space to compensate the twist of a torsioning group (such as methyl) so that the corresponding normal vibration generates zero overall angular momentum. But it has been shown that both mechanisms are required if the theory is to be invariant, not only to the choice of molecular origin, but also to the choice of local bond or group origins [45,5].

The bond polarizability theory of conventional Raman intensity is well-established [46,47]. The starting point is Placzek's approximation for the vibrational Raman transition polarizability at transparent frequencies [48]. On expanding the effective polarizability operator $\alpha_{\alpha\beta}(Q)$ in the normal vibrational coordinates Q_p, the transition polarizability becomes

$$\langle m_v | \alpha_{\alpha\beta}(Q) | n_v \rangle = (\alpha_{\alpha\beta})_0 \, \delta_{\alpha\beta} + \sum_p \left(\frac{\partial \alpha_{\alpha\beta}}{\partial Q_p}\right)_0 \langle m_v | Q_p | n_v \rangle + \ldots \tag{4.1}$$

where $|n_v\rangle$ and $|m_v\rangle$ denote the initial and final vibrational states. If $|n_v\rangle$ and $|m_v\rangle$ are the ground and first excited states $|p_0\rangle$ and $|p_1\rangle$ associated with the normal mode Q_p, then [49]

$$\langle p_1 | Q_p | p_0 \rangle = (\hbar/2\omega_p)^{1/2} \tag{4.2}$$

The amplitude of the fundamental Raman wave is therefore determined by the variation of the molecular polarizability with the corresponding normal vibrational coordinate and this is turn is calculated by way of the variation of the tensor with local internal coordinates. Internal vibrational coordinates s_q, such as local bond stretches, angle bends and torsions, are written as sums over the set of normal vibrational coordinates [49]

$$s_q = \sum_{p=1}^{3N-6} L_{qp} Q_p \tag{4.3}$$

the L-matrix elements being obtained from a normal coordinate analysis. Thus

$$\left(\frac{\partial \alpha_{\alpha\beta}}{\partial Q_p}\right)_0 = \sum_q \left(\frac{\partial \alpha_{\alpha\beta}}{\partial s_q} \frac{\partial s_q}{\partial Q_p}\right)_0 = \sum_q \left(\frac{\partial \alpha_{\alpha\beta}}{\partial s_q}\right)_0 L_{qp} \tag{4.4}$$

The molecular polarizability operator is now written as a sum of local bond polarizability operators

$$\alpha_{\alpha\beta}(Q) = \sum_i \alpha_{i_{\alpha\beta}}(Q) \tag{4.5}$$

so that

$$\langle p_1 | \alpha_{\alpha\beta}(Q) | p_0 \rangle = \left(\frac{\hbar}{2\omega_p}\right)^{1/2} \sum_i \sum_q \left(\frac{\partial \alpha_{i_{\alpha\beta}}}{\partial s_q}\right)_0 L_{qp} \tag{4.6}$$

The extension of the bond polarizability theory to ROA is based on the origin-dependence of $G'_{\alpha\beta}$ and $A_{\alpha\beta\gamma}$. Thus using (2.6) the optical activity tensors of the molecule, written as sums of corresponding bond tensors, are

$$G'_{\alpha\beta}(Q) = \sum_i G'_{i_{\alpha\beta}}(Q) - \frac{1}{2}\omega\varepsilon_{\beta\gamma\delta} \sum_i R_{i_\gamma}(Q)\, \alpha_{i_{\alpha\delta}}(Q) \tag{4.7 a}$$

$$\begin{aligned} A_{\alpha\beta\gamma}(Q) = {} & \sum_i A_{i_{\alpha\beta\gamma}}(Q) + \frac{3}{2} \sum_i R_{i_\beta}(Q)\, \alpha_{i_{\alpha\gamma}}(Q) \\ & + \frac{3}{2} \sum_i R_{i_\gamma}(Q)\, \alpha_{i_{\alpha\beta}}(Q) - \sum_i R_{i_\delta}(Q)\, \alpha_{i_{\alpha\delta}}(Q)\, \delta_{\beta\gamma} \end{aligned} \tag{4.7 b}$$

where $\mathbf{R}_i$ is the vector from the molecular origin to the origin of the ith bond. Expanding each term in the normal vibrational coordinates and using (4.6), the products that determine the Raman intensity and optical activity are found to be [3,5,45]

$$\begin{aligned} & \langle p_0 | \alpha_{\alpha\beta} | p_1 \rangle \langle p_1 | \alpha_{\alpha\beta} | p_0 \rangle \\ & = \frac{\hbar}{2\omega_p} \left[\sum_i \sum_q \left(\frac{\partial \alpha_{i_{\alpha\beta}}}{\partial s_q}\right)_0 L_{qp}\right] \left[\sum_j \sum_r \left(\frac{\partial \alpha_{j_{\alpha\beta}}}{\partial s_r}\right)_0 L_{rp}\right] \end{aligned} \tag{4.8 a}$$

$$\langle p_0 | \alpha_{\alpha\beta} | p_1 \rangle \langle p_1 | G'_{\alpha\beta} | p_0 \rangle$$

$$= -\frac{\hbar\omega}{4\omega_p} \varepsilon_{\beta\gamma\delta} \left\{ \sum_{i<j} (R_{ji_\gamma})_0 \left[\sum_q \left(\frac{\partial \alpha_{i_{\alpha\beta}}}{\partial s_q} \right)_0 L_{qp} \right] \left[\sum_r \left(\frac{\partial \alpha_{j_{\delta\alpha}}}{\partial s_r} \right)_0 L_{rp} \right] \right.$$
$$\left. + \left[\sum_i \sum_q \left(\frac{\partial \alpha_{i_{\alpha\beta}}}{\partial s_q} \right)_0 L_{qp} \right] \left[\sum_j (\alpha_{j_{\delta\alpha}})_0 \sum_r \left(\frac{\partial R_{j_\gamma}}{\partial s_r} \right)_0 L_{rp} \right] \right\}$$
$$+ \frac{\hbar}{2\omega_p} \left[\sum_i \sum_q \left(\frac{\partial \alpha_{i_{\alpha\beta}}}{\partial s_q} \right)_0 L_{qp} \right] \left[\sum_j \sum_r \left(\frac{\partial G'_{j_{\alpha\beta}}}{\partial s_r} \right)_0 L_{rp} \right] \quad (4.8\,b)$$

$$\frac{1}{3} \omega \langle p_0 | \alpha_{\alpha\beta} | p_1 \rangle \langle p_1 | \varepsilon_{\alpha\gamma\delta} A_{\gamma\delta\beta} | p_0 \rangle$$

$$= -\frac{\hbar\omega}{4\omega_p} \varepsilon_{\beta\gamma\delta} \left\{ \sum_{i<j} (R_{ji_\gamma})_0 \left[\sum_q \left(\frac{\partial \alpha_{i_{\alpha\beta}}}{\partial s_q} \right)_0 L_{qp} \right] \left[\sum_r \left(\frac{\partial \alpha_{j_{\delta\alpha}}}{\partial s_r} \right)_0 L_{rp} \right] \right.$$
$$\left. + \left[\sum_i \sum_q \left(\frac{\partial \alpha_{i_{\alpha\beta}}}{\partial s_q} \right)_0 L_{qp} \right] \left[\sum_j (\alpha_{j_{\delta\alpha}})_0 \sum_r \left(\frac{\partial R_{j_\gamma}}{\partial s_r} \right)_0 L_{rp} \right] \right\}$$
$$+ \frac{\hbar\omega}{6\omega_p} \left[\sum_i \sum_q \left(\frac{\partial \alpha_{i_{\alpha\beta}}}{\partial s_q} \right)_0 L_{qp} \right] \left[\varepsilon_{\alpha\gamma\delta} \sum_j \sum_r \left(\frac{\partial A_{j_{\gamma\delta\beta}}}{\partial s_r} \right)_0 L_{rp} \right] \quad (4.8\,c)$$

The first term in (4.8b and c) is a sum over all pairs of groups that constitute chiral structures, $(\mathbf{R}_{ji})_0 = (\mathbf{R}_j)_0 - (\mathbf{R}_i)_0$ being the vector from the origin on group i to that on j at the equilibrium nuclear configuration. This term is simply a generalization of the two-group mechanism [43]. The second term involves changes in the position vector $\mathbf{R}_j$ of a group relative to the molecular origin, and is a generalization of the inertial mechanism [44]. The last terms involve products of intrinsic group polarizability and optical activity tensors. As discussed elsewhere [45,5], these ROA expressions are invariant both to the choice of the molecular origin and the local group origins.

If a bond has a threefold for higher rotation axis, its polarizability tensor can be written as follows in terms of components referred to principal axes:

$$\alpha_{\alpha\beta} = \alpha_\perp \delta_{\alpha\beta} + \Delta u_\alpha u_\beta \quad (4.9)$$

where $\Delta = \alpha_{||} - \alpha_\perp$, $\alpha_{||}$ and $\alpha_\perp$ denoting components parallel and perpendicular to the bond axis, and $\mathbf{u}$ is the unit vector along the bond axis. The mean polarizability in this instance is

$$\alpha = \frac{1}{3} \alpha_{\alpha\alpha} = \frac{1}{3} (2\alpha_\perp + \alpha_{||})$$

As demonstrated elsewhere [45,5], if all the bonds in the molecule are axially-symmetric and achiral, the terms in the ROA expressions (4.8) involving intrinsic bond optical activity tensors $G'_{i_{\alpha\beta}}$ and $A_{i_{\alpha\beta\gamma}}$ vanish. For this particular case, the bond polarizability expressions (4.8) have been developed into a computational form that enables the ROA to be calculated for any normal model [45]; but because of their complexity, we shall not reproduce the explicit expressions here. The information necessary to initiate a general bond polarizability calculation falls into two categories. First there is the information from the normal coordinate analysis: Wilson's s-vectors $\mathbf{s}_m^t$ associated with each atom m and internal coordinate s_t, which are required to set up the vibrational **G**-matrix, and the **L** and $\mathbf{L}^{-1}$ matrices [49], all of which may be obtained routinely from a normal coordinate analysis program. The second type of information required is the bond polarizability data. Three parameters per bond i are needed: $\alpha'_{i\perp}$, the derivative with respect to the stretch coordinate of the perpendicular polarizability component, the bond polarizability anisotropy Δ_i and its derivative Δ'_i with respect to the stretch coordinate. Approximate values of Δ_i, deduced from Kerr effect and light scattering measurements, are available for most common bonds in various environments; whereas values of $\alpha'_{i\perp}$ and Δ'_i for a few bonds have been deduced from absolute Raman intensity studies.

Two detailed calculations of complete ROA spectra using this computational method have been reported:
(R)—(+) bromochlorofluoromethane [45] and (R)—(+) 3-methylcyclohexanone [50]. The calculated effects for bromochlorofluoromethane are rather small, with no CID Δ-value exceeding $\pm 1 \times 10^{-4}$, and are quite different in sign and magnitude to the effects calculated using the atom-dipole interaction theory discussed below. However, no ROA measurements have yet been reported on this molecule so it is not known which theory gives the better description. The calculated ROA spectrum of 3-methylcyclohexanone has been compared with the observed depolarized spectrum in the region 50–1800 cm^{-1}. The calculation is reasonably successful above about 1200 cm^{-1}, with most of the observed effects predicted with the correct sign; but below 1200 cm^{-1} the correlation is poor. Worse effects are expected at lower frequencies in this molecule for two reasons. First, the normal modes become much more complicated and the normal mode calculations less reliable. Second, the contribution to normal modes from internal coordinates involving carbonyl deformations become greater: the carbonyl group was incorporated into the calculation using the poor approximation of cylindrical symmetry, thereby losing contributions from the intrinsic optical activity tensors $G'_{\alpha\beta}$ and $A_{\alpha\beta\gamma}$ which probably dominate the large ROA associated with carbonyl deformations (see below).

4.2 The Atom Dipole Interaction Theory

The atom dipole interaction theory has quite a long history and has provided useful physical insight into optical properties such as optical rotation and the Kerr effect [51]. It was first applied to ROA by Prasad and Burow [52,53] and re-formulated by Prasad and Nafie [54]. In its application to ROA, the starting point is formally very similar to that of the bond polarizability theory, namely equations (4.7) expressing the optical activity tensors of the molecule as sums over the units that are considered to constitute a molecule. But the physical content is quite different because now these units are atoms, rather than bonds (so perhaps 'atom polarizability theory' would be a better name). As discussed at length elsewhere [3,5,45], dynamic coupling (which refers to the modification of the electromagnetic fields experienced at a second atom on account of the fields radiated by a first atom under the influence of the light wave) must be invoked at the outset, otherwise the molecule is completely isotropic and optically inactive.

Thus the polarizability tensor of the molecule is written as a sum of local atomic polarizabilities, each modified through dipolar interactions with the electric dipole moments on all the other atoms induced by the electric vector of the incident light wave. Similarly for the local atomic polarizabilities appearing in the origin-dependent parts of the optical activity tensors. But unlike the bond polarizability development, no allowance can be made for intrinsic local optical activity tensors $G'_{i_{\alpha\beta}}$ and $A_{i_{\alpha\beta\gamma}}$ since these now pertain to spherical atoms. We refer to the original articles for the explicit Raman intensity and optical activity expressions generated by the atom dipole interaction theory.

Apart from a normal coordinate analysis, rather less information is required to perform an atom dipole interaction calculation than a bond polarizability calculation of ROA: specifically, the derivatives of the local isotropic atomic polarizabilities with respect to the internal coordinates, which can be decuded from ordinary Raman intensities or else transferred from other (simpler) molecules. The anisotropic parts of the polarizability derivatives are generated automatically by the dipole-dipole interaction function.

This theory has now been applied to calculate the complete ROA spectra of a number of chiral molecules. The first calculation was for CHFClBr [53]: as mentioned above, the results are quite different to those from the bond polarizability theory, being consistently an order of magnitude larger and often of the opposite sign. Calculations on CDFClBr were also reported in the same paper. The next calculation to be reported was on the C—Cl stretching region of 1-chloro-2-methylbutane [55]. Agreement in sign and band contour was found, but the calculated ROA was an order of magnitude smaller than observed. However, the interpretation was obscured by uncertainties about the conformer populations. Mixed results were obtained for the 100–2000 cm^{-1} region of (R)—(+) 3-methylcyclohexanone [56]: a number of the prominent ROA features are predicted to have the correct sign, but magnitudes are unreliable. One notable success was the prediction of the large couplet observed at 942 and 968 cm^{-1} attributed to the orthogonal methyl rock modes. Calculations have also been reported on bromochlorofluoroethane and a deuterated analogue [57] as part of a study of the vibrational optical activity associated with 'perturbed degenerate modes' such as the methyl asymmetric deformations and the orthogonal methyl rockings: in most cases degenerate mode pairs gave calculated ROA couplets of opposite sign and almost equal magnitude.

4.3 Localized Molecular Orbital Theories

Localized molecular orbitals (LMOs) are certain combinations of delocalized molecular orbitals (eigenfunctions of the Hamiltonian) such that charge density is concentrated in particular regions of the molecule. Individual LMOs can therefore be identified with bond orbitals between a pair of atoms, lone pair orbitals on isolated atoms, and inner shell atomic orbitals. LMO methods have recently been introduced as a means of calculating ROA spectra without the necessity for extracting parameters from experimental intensity data [58-60]. The molecular polarizability is regarded as the sum of the polarizabilities of localized MOs.

Thus Eq. (4.7) for the optical activity tensors of the molecule are again employed, but now the summation is over all occupied LMOs and the vectors $\mathbf{R}_i$ define the positions of the orbital centroids. Once the wavefunctions are known, the polarizability $\alpha_{i_{\alpha\beta}}$ of the ith LMO and the position of its centroid $\mathbf{R}_i$ can be determined. The derivatives of $\alpha_{i_{\alpha\beta}}$ and $\mathbf{R}_i$ with respect to the normal coordinates are calculated using the electric field perturbation approach recently shown to be very effective for the calculation of conventional infrared and Raman intensities [61]: the required derivatives of $\mathbf{R}_i$ and $\alpha_{i_{\alpha\beta}}$ are determined from the first and second derivatives, respectively, of the gradient of the molecular potential energy with respect to a small applied electric field. One important aspect of this method is that both infrared CD and ROA can be determined from the same conceptual and calculational method, which will enhance the study of the relationship between these two forms of vibrational optical activity. So far, only one ROA calculation using LMO methods has been reported [59], and since that was for the model compound NHDT there has been no comparison with experimental data.

4.4 Simple Models

On account of the complexity of the normal modes in typical chiral molecules, the general computational theories are usually required to interpret observed ROA spectra. However, simple models can provide helpful physical insight into possible sources of ROA and can occasionally provide a realistic interpretation in bands originating in normal modes dominated by just one or two internal coordinates.

In an early and rather naive theory, symmetry rules were developed in terms of static non-bonded interactions with the rest of the molecule that induced electronic chirality into locally achiral groups on which characteristic vibrations are localized [62]. This aspect can be extended to include dynamic coupling, and further discussion can be found in Ref. 5. However, as discussed above, the vibrational chirality viewpoint has been found to be more fruitful in most cases.

One advantage of the bond polarizability theory is that, since it is based on a decomposition of the molecule into bonds or groups that can support local internal vibrational coordinates, it can be applied to idealized normal modes containing just a few internal coordinates and so can provide conceptual models of the generation of ROA by some simple chiral structures. Indeed, as mentioned above, the bond polarizability theory actually developed out of a synthesis of the two-group model and the inertial model, both of which have been applied in detail to a number of simple chiral structures [3,5].

Consider, for example, a simple chiral two-group structure where the principal axes of two axially-symmetric neutral equivalent achiral groups 1 and 2 are in parallel planes. Origins lying along the principal axes of the two groups are connected by a vector $\mathbf{R}_{21} = \mathbf{R}_2 - \mathbf{R}_1$, and θ is the torsion angle between the two principal axes. Idealized normal coordinates containing symmetric and antisymmetric combinations of two equivalent internal coordinates s_1 and s_2 (such as bond stretches) localized on groups 1 and 2 can be written

$$Q_+ = N(s_1 + s_2) \quad (4.10a)$$

$$Q_- = N(s_1 - s_2) \quad (4.10b)$$

where N is a constant. Application of the first terms of (4.8b and c) for the two-group optical activity, together with (4.8a) for the intensity, gives the following depolarized Raman CIDs for the fundamental transitions in the two normal modes[43, 3, 5].

$$\Delta_z(1_+ \leftarrow 0) = \frac{2\pi R_{21} \sin 2\theta}{\lambda(5 + 3 \cos 2\theta)} \quad (4.11a)$$

$$\Delta_z(1_- \leftarrow 0) = \frac{-2\pi R_{21} \sin 2\theta}{3\lambda(1 - \cos 2\theta)} \quad (4.11b)$$

Although the dimensionless CIDs have different magnitudes and opposite signs for Q_+ and Q_-, $I_z^R - I_z^L$ itself has equal magnitudes and opposite signs and so generates a conservative ROA couplet. The CIDs associated with other modes of this structure, such as the torsion and deformations, have also been calculated[5]. Other aspects of the Rayleigh and Raman optical activity of the two-group structure have been discussed, including extensions to more general geometries[2, 63, 64] and to group separations R_{21} much greater than the wavelength of the incident light[2, 5, 65, 66]. The only reported example of a detailed application of the simple two-group model to observed ROA features is to CH_2 scissoring vibrations in terpenes containing only two methylene groups[67] (see below).

Turning to the inertial model, this was originally exemplified in a 'hindered single-bladed propellor' structure[44, 3, 5] in which a methyl group is attached to a contrived chiral framework such that the methyl torsion axis is a principal inertial axis of the molecule. Thus an anisotropic achiral group is oriented relative to the threefold axis of the methyl group such that the anisotropic group and the threefold axis constitute a chiral structure, the anisotropic group being balanced dynamically by a spherical group. The depolarized CID associated with the fundamental methyl torsion transition is found to be

$$\Delta_z(1_t \leftarrow 0) = \frac{2\pi R \sin 2\varphi \sin \theta}{3\lambda(1 - \cos 2\varphi)} \quad (4.12)$$

where R, θ and φ define the moment arm and orientation of the counter-oscillating group. Chiral molecules containing a single methyl group with its threefold axis lying along a principal inertial axis are rare (possible non-existent!), but the model can be extended to a common situation in which the symmetric combination of the torsions of two adjacent methyl groups can generate oscillations of the complete molecule about a twofold rotation axis, but now two-group terms usually have to be included as well. An example of such a calculation has recently been reported for 2,3-epoxybutane [68], good agreement being found with experiment (see below).

Finally, there are the terms in the bond polarizability ROA expressions (4.8) involving intrinsic bond or group optical activity tensors $G'_{i_{\alpha\beta}}$ and $A_{i_{\alpha\beta\gamma}}$. In general, these will survive for bonds or groups lacking a threefold or higher rotation axis, and for non-totally symmetric internal vibrational coordinates that destroy the axial symmetry of axially-symmetric groups. Consider two idealized normal coordinates containing symmetric and antisymmetric combinations of two internal coordinates, in general non-equivalent, localized on the same group i:

$$Q_+ = N_1 s_1 + N_2 s_2 \tag{4.13a}$$

$$Q_- = N_2 s_1 - N_1 s_2 \tag{4.13b}$$

The required contribution to (4.8b) is

$$\langle 0 | \alpha_{i_{\alpha\beta}} | 1_\pm \rangle \langle 1_\pm | G'_{i_{\alpha\beta}} | 0 \rangle$$
$$= \pm \left(\frac{\hbar}{2\omega_\pm}\right) \frac{N_1 N_2}{(N_1^2 + N_2^2)^2} \left[\left(\frac{\partial \alpha_{i_{\alpha\beta}}}{\partial s_1}\right)_0 \left(\frac{\partial G'_{i_{\alpha\beta}}}{\partial s_2}\right)_0 + \left(\frac{\partial \alpha_{i_{\alpha\beta}}}{\partial s_2}\right)_0 \left(\frac{\partial G'_{i_{\alpha\beta}}}{\partial s_1}\right)_0 \right] \tag{4.14}$$

with an analogous contribution to (4.8c). A possible example is the carbonyl group in molecules such as 3-methylcyclohexanone. The in-plane and out-of-plane deformation coordinates belong to symmetry species B_2 and B_1 in the local C_{2v} symmetry: B_2 is spanned by α_{YZ}, G'_{XZ} and G'_{ZX}; and B_1 by α_{XZ}, G'_{YZ} and G'_{ZY}. The skeletal chirality will lead to normal modes of vibration of the form (4.13), generating equal and opposite ROA. A simple analysis reduces (4.14) to [5,69]

$$\langle 0 | \alpha_{\alpha\beta} | 1_\pm \rangle \langle 1_\pm | G'_{\alpha\beta} | 0 \rangle$$
$$= \pm \left(\frac{\hbar}{2\omega_\pm}\right) \frac{N_1 N_2}{(N_1^2 + N_2^2)^2} (2\alpha_{ZZ} - \alpha_{XX} - \alpha_{YY})(G'_{XY} + G'_{YX}) \tag{4.15}$$

Although greatly over-simplified, such a mechanism probably has some rôle in the generation of the large ROA couplets that often appear around 500 cm^{-1} in chiral molecules containing carbonyl groups attached to a ring system [69] (see below). Notice that similar deformations of an axially-symmetric group can generate no corresponding ROA because now $G'_{XY} = -G'_{YX}$ [5].

4.5 Sum Rules

Sum rules in ROA (and indeed in infrared CD) constitute an important unsolved problem at present. One aspect would involved statements concerning the Rayleigh optical activity, or the optical activity ($I^R - I^L$) associated with a given Raman band, when integrated over all incident frequencies from zero to infinity (the result might be zero). The other aspect, of greater practical significance, is the value of $I^R - I^L$ summed over a complete vibrational Raman spectrum for a fixed incident frequency: the result is not expected to be zero, but it might be related to the value for the Rayleigh line (this is intimated by the fact that the 'conservative couplets' mentioned above in connection with the two-group model are not strictly conservative since $I^R - I^L$ for Q_+ and Q_- is a function of ω_+^{-1} and ω_-^{-1} which are slightly different). Also, it should be remembered that any general sum rule should include electronic and rotational Raman transitions as well as vibrational, and that vibrational-electronic coupling is required for vibrational Raman scattering at transparent frequences.

Cuony and Hug [70] and Hug [38] have discussed sum rules for vibrational ROA in the limited context of molecules that are chiral due to isotopic substitution. By expressing the normal coordinates in $(\partial\alpha_{\alpha\beta}/\partial Q_p)_0$ etc. in terms of atomic cartesian displacement vectors, they were able to show that the optical activity in the fundamentals sums to zero if the parent achiral group is other than C_{nh} and C_i. Their proof requires the assumption that the isotopic substitution does not affect the equilibrium electronic distribution, in which case the corresponding Rayleigh optical activity is zero.

5 Experimental Results

A large number of ROA spectra have now been measured and discussed. Most have been obtained with the scanning instrument in Glasgow and so are depolarized and cover the region from about 80 to 1800 cm^{-1}. However, a sizeable number of multichannel spectra from the Toronto and Fribourg instruments have also been published, and these are usually of better quality: in particular, the Fribourg spectra sometimes include the weaker effects in the 2000–3500 cm^{-1} range, as well as one or two polarized spectra. Apart from the different sign convention in the published spectra from the different groups, another difference is that some spectra are presented on a linear scale, whereas others are presented on a scale that is linear within each decade range but logarithmic between decade ranges (see above).

Figure 6 shows a good example of the spectra obtained with the Glasgow scanning instrument, that of (—) menthol in methanol solution. The complicated ROA structure between about 1100 and 1400 cm^{-1} is repeated almost exactly (in sign and magnitude) in (—)menthylamine and (—)menthyl chloride [71], which illustrates how valuable such comparisons can be for correlating stereochemical features such as absolute configuration in series of related compounds. This paper also showed the ROA spectra of (—)isopulegol, (—)menthone, (—)3,3-dimethylcyclohexanol, (+)pulegone, (—)limonene and (+)carvone, and pointed out several stereochemical correlations involving, in particular, bonds characteristic of the isopropyl group, the *gem*-dimethyl group, methyl torsions and out-of-plane olefinic hydrogen deformations. In a related paper [72], the ROA spectra of (+) and (—)α-pinene, (—)β-pinene, (—)cedrene, (+)car-3-ene, (+)car-2-ene, (—)caryophyllene and (—)β-bourbonene were presented, and a stereochemical correlation in bonds originating in skeletal modes of the two pinenes pointed out as well as further examples of

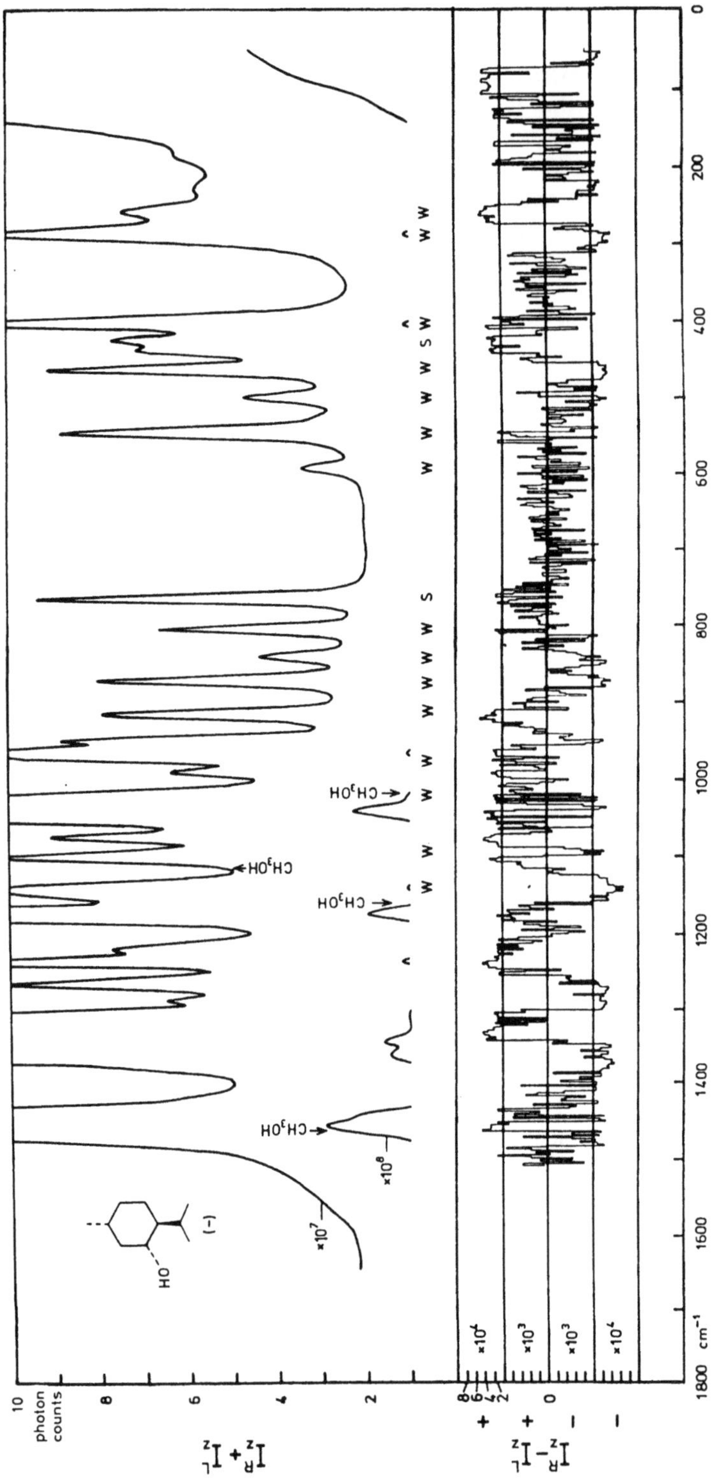

Fig. 6. The depolarized Raman circular intensity sum $(I_z^R + I_z^L)$ and difference $(I_z^R - I_z^L)$ spectra of (−) menthol in methanol solution

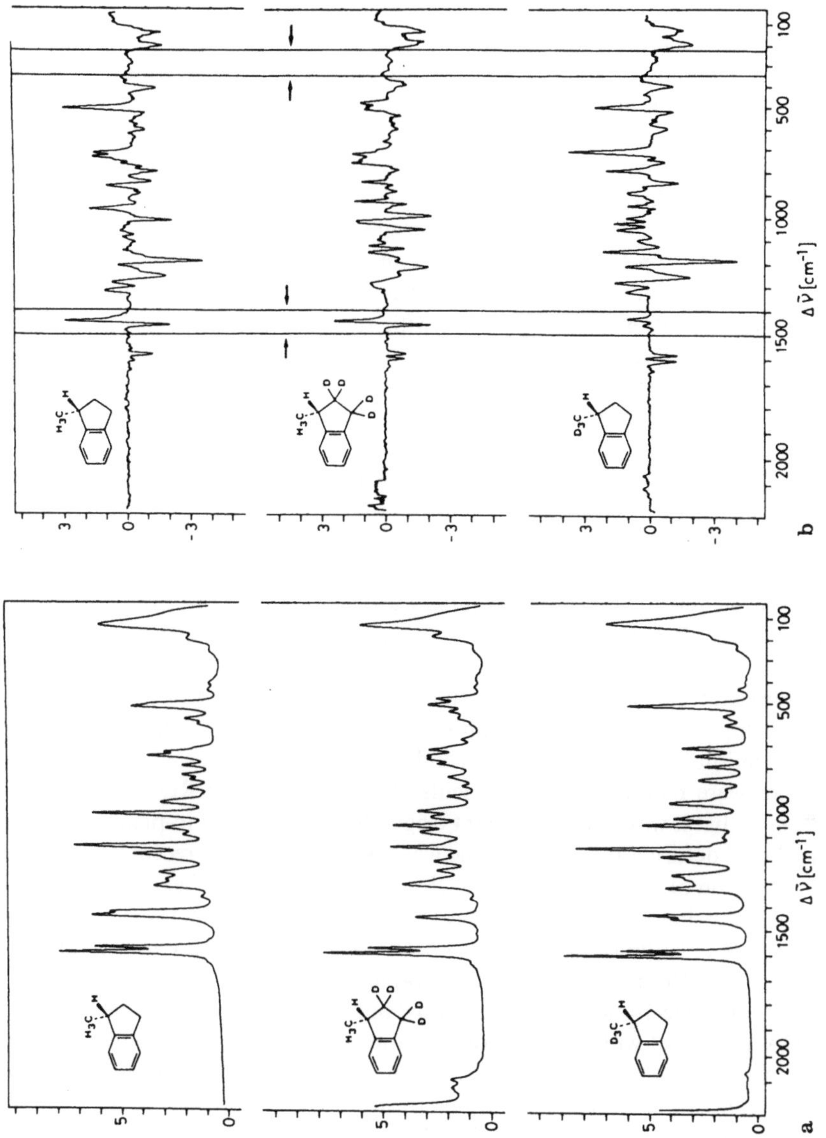

Fig. 7. (a) the depolarized Raman spectra $(I_z^L + I_z^R)/2$ and (b) the circular difference spectra 1000 $(I_z^L - I_z^R)$ of neat (R)—(+) 1-methylindane and two deuterated derivatives. The chirality numbers q times a factor of 1000 are obtained by dividing the spectra in (b) by the corresponding ones from (a). Reproduced from Ref. 76 with the publisher's permission

effects in characteristic bands of the *gem*-dimethyl group and the out-of-plane olefinic hydrogen deformations.

The first observations of ROA were made in α-phenylethylamine and α-phenylethanol in the form of a couplet centered at about 340 cm^{-1} [18]. Subsequent measurements on these and related arylethanes [21,73,74] revealed a sharp couplet at about 1450 cm^{-1} associated with the methyl asymmetric deformations. At first it was thought that this couplet originated in the two degenerate components of the methyl asymmetric deformations being slightly split by the chiral environment [21,73]. However, the α-p-bromophenylethylamine shows no such couplet, which indicates that the lower-frequency semicircle stretching mode of the aromatic ring is involved in the generation of the couplet, with the corresponding normal modes embracing the highly chiral Ar—C*—CH_3 structure [74]. This conclusion is reinforced by parallel infrared CD studies [75], and by a definitive ROA study of 1-methylindane and several deuterated derivatives by Hug et al. [76] whose multichannel ROA spectra are shown in Fig. 7. It should be mentioned that although the original description of the origin of the methyl asymmetric deformation ROA couplet proved to be over-simplified, it might still be valid for the trifluoromethyl group since a conservative couplet appears at 510 cm^{-1} in α-phenyltrifluoromethylethanol [74], this region being appropriate for trifluoromethyl asymmetric deformations. Also, the spectra of (+) and (—)α-phenylethylisocyanate in ref. 19 were assigned to the wrong enantiomers, and this has since been corrected in a paper that presents a more complete spectrum [77].

Another series of molecules that was studied early on consisted of (+)camphor, (+)3-bromocamphor, (—)3-bromocamphor-9-sulphonic acid, (+)3-chlorocamphor, (—)bornan-2-endo-ol, (—)bornan-2-exo-ol and (+)3-methylcyclohexanone [78]. Several correlations were pointed out, in particular large conservative ROA couplets at about 500 cm^{-1} that it was suggested might originate in coupling of in-plane and out-of-plane carbonyl deformations. This theme was taken up in a more recent work [69] in which a comparison of the ROA spectra of (R)—(+)3-methylcyclohexanone (Fig. 8) and R—(—)3-methylmethylene cyclohexane (Fig. 9) provides convincing evidence for the dominant role of carbonyl deformations in the generation of this couplet, further evidence being provided by the spectra of (—)β-pinene, (+)nopinone and (R)—(—)-5-methylcyclohex-2-ene-1-one. However, it now appears that the most likely source of large ROA associated with carbonyl deformations is not a simple coupling of the two orthogonal deformations, but rather normal modes empbracing a highly chiral structure and containing large contributions from one of the carbonyl deformation coordinates and, for example, a C—C—Me deformation coordinate [79].

The ROA spectra of tartaric acid, dimethyl tartrate, 2,3-butanediol and dibenzoyl tartaric acid have been reported and discussed [80]. One striking feature in all these spectra is a large couplet at about 500 cm^{-1} that might originate in deformations of the twisted O—C—C—O unit.

A series of chiral sulphoxides, namely the p-tolylethyl, p-tolylmethyl, p-tolyl-isopropyl, p-tolyl-*tert*-butyl and p-tolyl-*o*-tolyl, has been studied [23]. All showed large ROA below about 500 cm^{-1}, especially in bands assigned to C—S—O deformations. It appears to be a general feature of both infrared CD and ROA that the presence of sulphur atoms greatly enhances the vibrational optical activity.

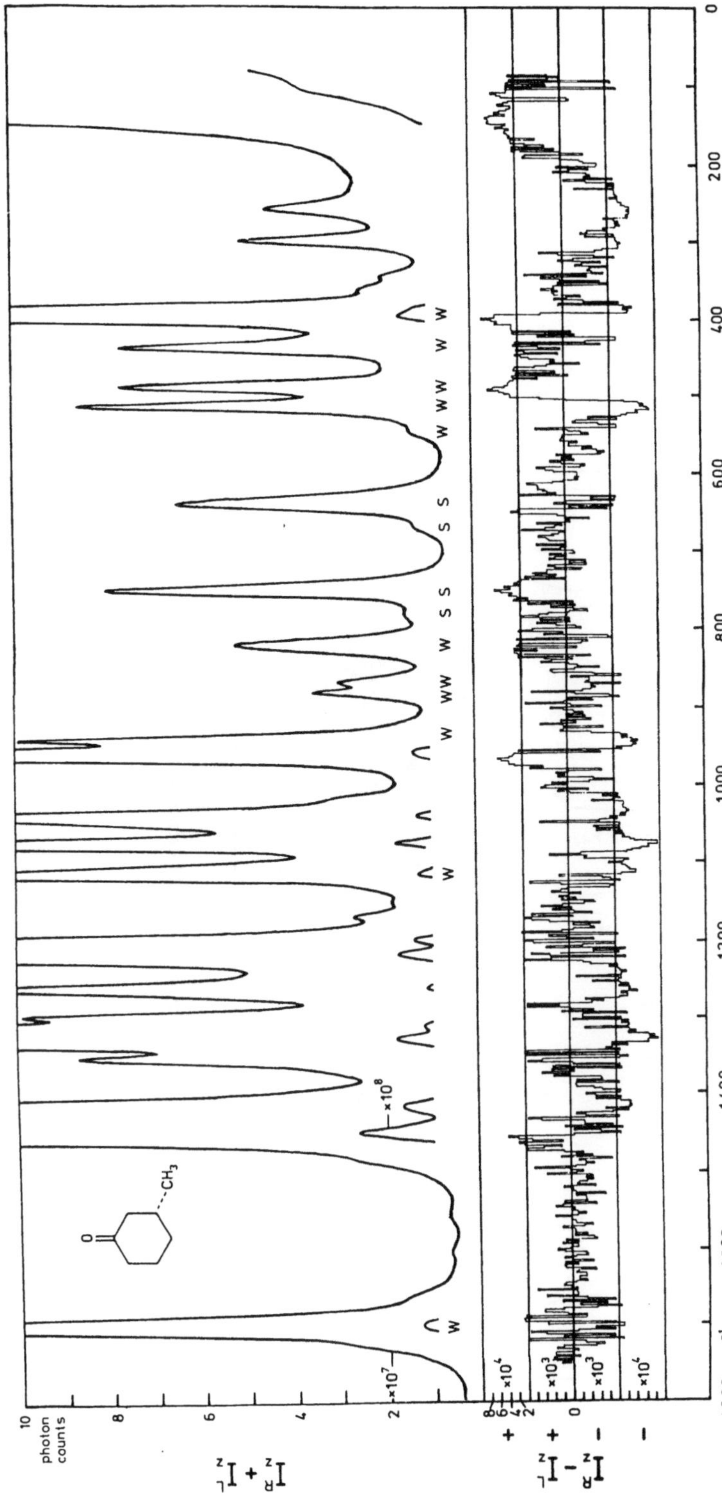

Fig. 8. The depolarized Raman circular intensity sum and difference spectra of neat (R)—(+) 3-methylcyclohexanone

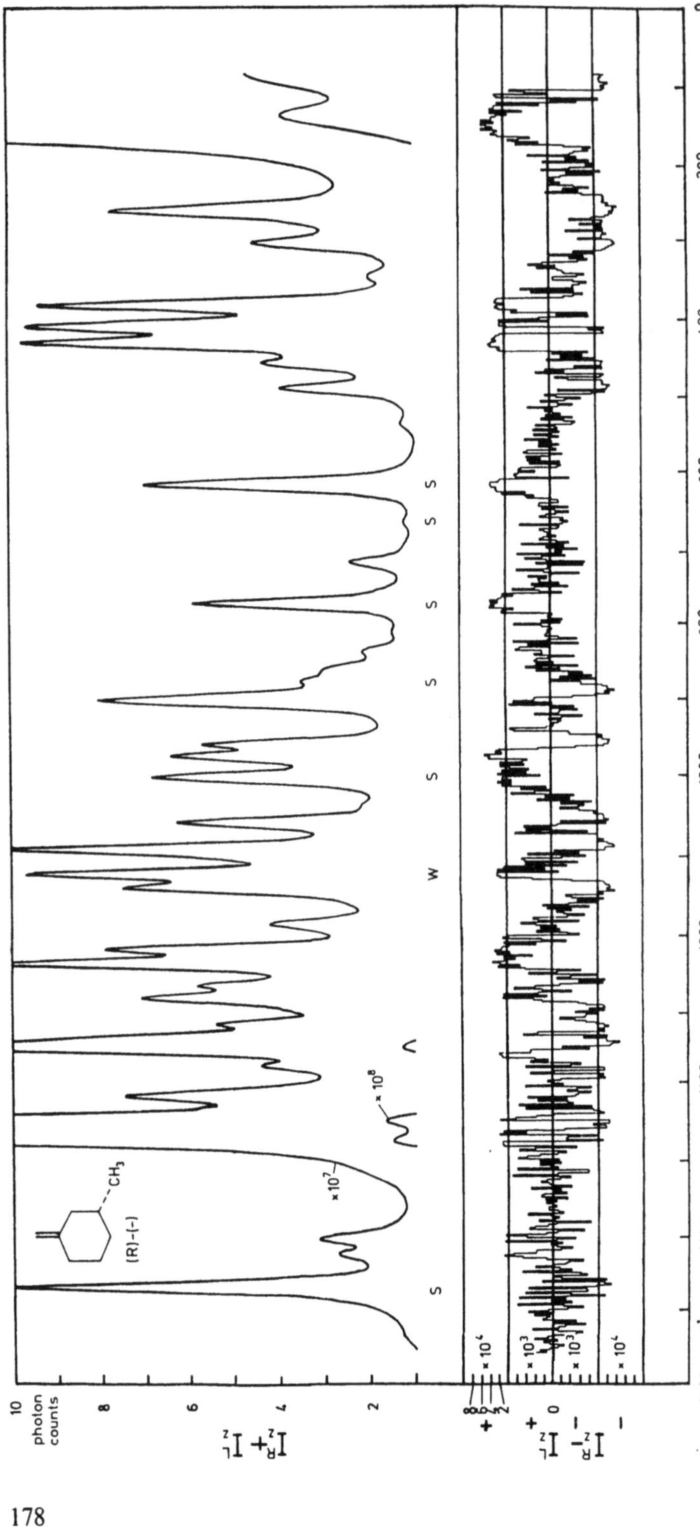

Fig. 9. The depolarized Raman circular intensity sum and difference spectra of neat (R)—(—) 3-methyl methylenecyclohexane

The Toronto multichannel instrument was used to record the ROA spectra of a series of ten chiral terpenes, namely *cis*-pinane, *cis*-myrtanol, *cis*-myrtanylamine, α-pinene, myrtenol, nopol, myrtenal, *trans*-verbenol, β-pinene and *cis*-3-pinen-2-ol [34]. A number of common features were noted which it was suggested may form the basis of an absolute configuration rule for molecules belonging to this series. A particularly interesting observation was that a ROA couplet appeared in the CH_2 scissoring region in the spectra of most of these molecules. This was attributed, plausibly, to a two-group mechanism involving pairs of CH_2 groups that constitute highly chiral structures, and this theme was developed in detail in a subsequent paper [67]. In an independent joint ROA and infrared CD study of p-menth-1-ene and p-menth-1-en-9-ol [81], ring methylene modes, especially the scissoring, were again implicated as a major source of vibrational optical activity. Contributions from CH_2 scissoring modes were also isolated in the previously mentioned study of deuterated 1-methyl indanes [76].

A few ROA spectra of molecules that are chiral due to deuterium substitution have been reported. α-D-benzyl alcohol shows a couplet in two Raman bands at about 860 and 950 cm^{-1} that are not present in benzyl alcohol itself and which have been assigned to C—D deformations [82], although aromatic modes are probably involved as well. In another example, rich ROA structure was observed between about 800 and 1300 cm^{-1} in 4,4-dideuteroadamantan-2-one [83], this being the region for CH_2 and CD_2 deformations. Hug [38] has recently published a preliminary report of the multichannel ROA spectra of some deuterated cyclohexanones, and has discussed them in terms of sum rules.

There has been some discussion of the possible role of methyl torsions in the generation of Raman optical activity at low frequency (via the inertial mechanism discussed in Section 4.4 above) [73,44]. However, a comparison of the Raman and ROA spectra of 1-methylindane and the corresponding deutero methylindane (Fig. 7) provides no evidence for the involvement of methyl torsions in these molecules, at least above 220 cm^{-1}, and it was suggested that the concept of methyl torsion ROA could be dismissed, at least in arylethanes [76]. On the other hand, 2,3-epoxybutane shows large ROA in a band that has been independently assigned to the in-phase methyl torsion, and an application of the general bond polarizability theory provides a result that is in reasonable agreement with the observed value [84].

The ability of ROA to measure vibrational optical activity in very low frequency vibrations has important implications for the study of normal modes corresponding to the 'foothills' of the interconversion pathways between conformers. This is born out by observations of dramatic changes with temperature in the ROA of two low-frequency bands of α-phellandrene, which could be associated with the interconversion of the pseudoaxial and pseudoequatorial conformers [85].

The first reported ROA spectra from the scanning instrument in Tokyo involved a few of the bands of α-pinene, α-phenylethylamine, limonene and 3-carene, the paper being concerned mainly with checking the validity of the results [24]. A second paper reported ROA spectra of 2-butanol, 2-pentanol and 2-octanol and pointed out several correlations in the 700–1000 cm^{-1} region that were assigned to the symmetric stretching of the C_3O skeleton [86]. Recently, they have reported intriguing observations of *resonance* ROA arising from chirality induced in methyl orange molecules by inclusion in chiral cavities within α-cyclodextrin [87].

6 Concluding Remarks

It is now clear that ROA is a valuable new technique, not only for providing detailed stereochemical information, but also for fundamental studies of molecular vibrations in large low symmetry molecules. Now that there is a considerable overlap in the spectral regions accessible to both infrared CD and ROA, both these methods of studying vibrational optical activity will benefit greatly from detailed comparisons.

The problems associated with reliable, rapid, recording of ROA spectra have now been largely overcome by the use of optical mutichannel techniques in conjunction with a detailed theoretical understanding of the origin of the dominant artefacts. However, a simple scanning ROA spectrometer can still be of value for studying limited spectral regions of a favourable sample.

Considerable progress has been made in the development of theories that can predict the complete ROA spectrum, provided that a good normal coordinate analysis is available, and this leads us to the hope that it might be possible eventually to deduce the total stereochemistry (absolute configuration, conformation, bond lengths and angles) of a chiral molecule in a chemically relevant environment from the measured ROA spectrum (or indeed from the infrared CD spectrum).

One new field of application which the greatly increased sensitivity of optical multichannel techniques has opened up is to biological molecules, through both transparent and resonance Raman scattering. Infrared CD is likely to be inapplicable to most biological samples in aqueous media, so it is in biochemistry and biophysics that ROA will probably come into its own.

7 Acknowledgements

We thank the Science and Engineering Research Council for their support of the experimental ROA work carried out in Glasgow.

8 References

1. Barron, L. D., Buckingham, A. D.: Ann. Rev. Phys. Chem. *26*, 381 (1975)
2. Barron, L. D.: Raman optical activity, in: Advances in Infrared and Raman Spectroscopy, Vol. 4 (eds. Clark, R. J. H., Hester, R. E.) p. 271, London, Heyden 1978
3. Barron, L. D.: Raman optical activity, in: Optical Activity and Chiral Discrimination (ed. Mason, S. F.) p. 219, Dordrecht, Reidel 1979
4. Barron, L. D.: Acc. Chem. Res. *13*, 90 (1980)
5. Barron, L. D.: Molecular Light Scattering and Optical Activity, Cambridge, Cambridge Univ. Press 1982
6. Nafie, L. A., Diem, M.: Acc. Chem. Res. *12*, 296 (1979)
7. Nafie, L. A.: Infrared and Raman vibrational optical activity, in: Vibrational Spectra and Structure, Vol. 10 (ed. Durig, J. R.) p. 153, Amsterdam, Elsevier 1981
8. Polavarapu, P. L.: Recent advances in model calculation of vibrational optical activity, in: Vibrational Spectra and Structure, Vol. 13 (ed. Durig, J. R.), Amsterdam, Elsevier (in the press)
9. Gans, R.: Z. Phys. *17*, 353 (1923)
10. de Malleman, R.: Compt. Rend. *181*, 371 (1925)
11. Bhagavantam, S., Venkateswaran, S.: Nature *125*, 237 (1930)
12. Kastler, A.: Compt. Rend. *191*, 565 (1930)

13. Perrin, F.: J. Chem. Phys. *10*, 415 (1942)
14. Atkins, P. W., Barron, L. D.: Mol. Phys. *16*, 453 (1969)
15. Barron, L. D., Buckingham, A. D.: ibid. *20*, 1111 (1971)
16. Bosnich, B., Moskovits, M., Ozin, G. A.: J. Am. Chem. Soc., *94*, 4750 (1972)
17. Diem, M., Fry, J. L., Burow, D. F.: ibid. *95*, 253 (1973)
18. Barron, L. D., Bogaard, M. P., Buckingham, A. D.: ibid. *95*, 603 (1973)
19. Barron, L. D., Bogaard, M. P., Buckingham, A. D.: Nature *241*, 113 (1973)
20. Barron, L. D., Buckingham, A. D.: J. Chem. Soc., Chem. Commun. *1974*, 152
21. Hug, W., Kint, S., Bailey, G. F., Scherer, J. R.: J. Am. Chem. Soc. *97*, 5589 (1975)
22. Diem, M., Diem, M. J., Hudgens, B. A., Fry, J. L., Burow, D. F.: J. Chem. Soc., Chem. Commun. *1976*, 1028
23. Boucher, H., Brocki, T. R., Moskovits, M., Bosnich, B.: J. Am. Chem. Soc. *99*, 6870 (1977)
24. Waki, H., Higuchi, S., Tanaka, S., Sakayanag, N.: J. Spectrosc. Soc. Japan *26*, 272 (1977)
25. Barron, L. D., Vrbancich, J.: Magnetic Raman optical activity, in: Advances in Infrared and Raman Spectroscopy, (eds. Clark, R. J. H., Hester, R. E.) London, Heyden, in the press
26. Buckingham, A. D.: Adv. Chem. Phys. *12*, 107 (1967)
27. Long, D. A.: Raman Spectroscopy, New York, McGraw-Hill 1977
28. Andrews, D. L.: J. Chem. Phys. *72*, 4141 (1980)
29. Bjarnason, J. O., Andersen, H. C., Hudson, B. S.: ibid. *72*, 4132 (1980)
30. Oudar, J. L., Minot, C., Garetz, B. A.: ibid. *76*, 2227 (1982)
31. Barron, L. D.: Rayleigh and Raman scattering of polarized light, in: Molecular Spectroscopy, Vol. 4 (eds. Barrow, R. F., Long, D. A., Sheridan, J.) p. 96, London, The Chemical Society 1976
32. Buckingham, A. D., Longuet-Higgins, H. C.: Mol. Phys. *14*, 63 (1968)
33. Polavarapu, P. L., Diem. M., Nafie, L. A.: J. Am. Chem. Soc. *102*, 5449 (1980)
34. Brocki, T., Moskovits, M., Bosnich, B.: ibid. *102*, 495 (1980)
35. Hug, W., Surbeck, H.: Chem. Phys. Letters *60*, 186 (1979)
36. Talmi, Y.: Optoelectronic image detectors in chemistry, in: Multichannel Image Detectors (ed. Talmi, Y.), A.C.S. Symposium series, Vol. 102, p. 3, Washington D.C., Americal Chemical Society 1979
37. Horlick, G.: Appl. Spectrosc. *30*, 113 (1976)
38. Hug, W.: Instrumental and theoretical advances in Raman optical activity, in: Raman Spectroscopy (eds. Lascombe, J., Huong, P. V.) p. 3, Chichester, Wiley, 1982
39. Barron, L. D., Torrance, J. F.: Chem. Phys. Letters (in the press)
40. Barron, L. D., Buckingham, A. D.: U.K. Pat. No. 6454 (1971); U.S. Pat. No. 3817634 (1974)
41. Hug, W.: Appl. Spectrosc. *35*, 115 (1981)
42. Barron, L. D., Vrbancich, J.: J. Raman Spectrosc. (in the press)
43. Barron, L. D., Buckingham, A. D.: J. Am. Chem. Soc. *96*, 4769 (1974)
44. Barron, L. D., Buckingham, A. D.: ibid. *101*, 1979 (1979)
45. Barron, L. D., Clark, B. P.: Mol. Phys. *46*, 839 (1982)
46. Sverdlov, L. M., Kovner, M. A., Krainov, E. P.: Vibrational Spectra of Polyatomic Molecules, Jerusalem, Israel Program for Scientific Translations 1974
47. Gussoni, M.: Infrared and Raman intensities from electro-optical parameters, in: Advances in Infrared and Raman Spectroscopy, Vol. 6 (eds. Clark, R. J. H., Hester, R. E.) p. 61, London, Heyden, 1980
48. Placzek, G.: Rayleigh and Raman scattering, in: Handbuch der Radiologie, Vol. 6, Part 2, (ed. Marx, E.) p. 205, Leipzig, Akademische Verlagsgesellschaft 1934
49. Wilson, E. B., Decius, J. C., Cross, P. C.: Molecular Vibrations, New York, McGraw-Hill 1955
50. Barron, L. D., Clark, B. P.: J. Raman Spectrosc. *13*, 155 (1982)
51. Applequist, J.: Acc. Chem. Res. *10*, 79 (1977)
52. Prasad, P. L., Burow, D. F.: J. Am. Chem. Soc. *101*, 800 (1979)
53. Prasad, P. L., Burow, D. F.: ibid. *101*, 806 (1979)
54. Prasad, P. L., Nafie, L. A.: J. Chem. Phys. *70*, 5582 (1979)
55. Prasad, P. L., Nafie, L. A., Burow, D. F.: J. Raman Spectrosc. *8*, 255 (1979)
56. Polavarapu, P. L., Nafie, L. A.: J. Chem. Phys. *73*, 1567 (1980)
57. Nafie, L. A., Polavarapu, P. L., Diem, M.: ibid. *73*, 3530 (1980)
58. Nafie, L. A., Freedman, T. B.: ibid. *75*, 4847 (1981)
59. Polavarapu, P. L.: ibid. *77*, 2273 (1982)

60. Freedman, T. B., Nafie, L. A.: ibid. *78*, 27 (1983)
61. Komornicki, A., McIver, J. W.: ibid. *70*, 2014 (1979)
62. Barron, L. D.: J. Chem. Soc. A. *1971*, 2899
63. Stone, A. J.: Mol. Phys. *29*, 1461 (1975)
64. Stone, A. J.: ibid. *33*, 293 (1977)
65. Andrews, D. L., Thirunamachandran, T.: Proc. Roy. Soc. *A358*, 297 (1977)
66. Andrews, D. L., Thirunamachandran, T.: ibid. *A358*, 311 (1977)
67. Gohin, A., Moskovits, M.: J. Am. Chem. Soc. *10*, 1660 (1981)
68. Barron, L. D., Vrbancich, J.: Mol. Phys. *48*, 833 (1983)
69. Barron, L. D., Torrance, J. F., Vrbancich, J.: J. Raman Spectrosc. *13*, 171 (1982)
70. Cuony, B., Hug, W.: Chem. Phys. Letters *84*, 131 (1981)
71. Barron, L. D., Clark, B. P.: J. Chem. Soc., Perkin Trans. II *1979*, 1164
72. Barron, L. D., Clark, B. P.: ibid. *1979*, 1171
73. Barron, L. D.: Nature *255*, 458 (1975)
74. Barron, L. D.: J. Chem. Soc., Perkin Trans. II *1977*, 1790
75. Su, C. N., Keiderling, T. A.: Chem. Phys. Letters *77*, 494 (1981)
76. Hug, W., Kamatari, A., Srinivasan, K., Hansen, H.-J., Sliwka, H.-R.: ibid. *76*, 469 (1980)
77. Barron, L. D., Clark, B. P.: J. Chem. Res. (S) *1979*, 36
78. Barron, L. D.: J. Chem. Soc., Perkin Trans II *1977*, 1074
79. Torrance, J. F.: Ph. D. Thesis, Glasgow Univ. 1983
80. Barron, L. D.: Tetrahedron *34*, 607 (1978)
81. Polavarapu, P. L., Diem, M., Nafie, L. A.: J. Am Chem. Soc. *102*, 5449 (1980)
82. Barron, L. D.: J. Chem. Soc., Chem. Commun. *1977*, 305
83. Barron, L. D., Numan, H., Wynberg, H.: ibid. *1978*, 259
84. Barron, L. D., Vrbancich, J.: Mol. Phys. *48*, 833 (1983)
85. Barron, L. D., Vrbancich, J.: J. Chem. Soc., Chem. Commun. *1981*, 771
86. Waki, H., Higuchi, S., Tanaka, S.: Chem. Letters *1981*, 147
87. Higuchi, S., Tanaka, K., Tanaka, S.: ibid. *1982*, 635

Author Index Volumes 101–123

Contents of Vols. 50–100 see Vol. 100
Author and Subject Index Vols. 26–50 see Vol. 50

The volume numbers are printed in italics

Bei Fragen zur Produktsicherheit wenden Sie sich bitte an:
If you have any questions regarding product safety,
please contact:

Walter de Gruyter GmbH
Genthiner Straße 13
10785 Berlin
productsafety@degruyterbrill.com